Principles and Practices of Agricultural Disaster Management

Principles and Practices of
Agricultural Disaster Management

Dr. V. Radha Krishna Murthy
Ph.D, PGDES
Professor (Academic), O/o Dean of Agriculture
Acharya N. G. Ranga Agricultural University
Rajendranagar, Hyderabad – 30.

BSP **BS Publications**

A unit of **BSP Books Pvt. Ltd.**

4-4-309/316, Giriraj Lane, Sultan Bazar,
Hyderabad - 500 095
Phone : 040 - 23445605, 23445688

Principles and Practices of Agricultural Disaster Management
by V. Radha Krishna Murthy

© 2016 by publisher

Published by

BSP BS Publications

A unit of **BSP Books Pvt. Ltd.**

4-4-309/316, Giriraj Lane, Sultan Bazar,
Hyderabad - 500 095
Phone : 040 - 23445605, 23445688
e-mail : info@bspbooks.net
www.bspbooks.net

ISBN: 978-93-52301-06-5 (HB)

Dedicated to,

Lord Sri Tirumala Tirupati Venkateswara Swamy Vaariki

Dr. A. Padma Raju
Vice-Chancellor

Acharya N. G. Ranga Agricultural University
Rajendranagar, Hyderabad - 500 030.
Phone : 040- 24015035 (O)
040-24015031 (F)
Email : angrau_vc@yahoo.com
Grams: "AGRIVARSTY"

Foreword

Agriculture plays a significant role in addressing poverty, hunger, malnutrition and livelihood security of millions of people. However, crop production is highly dependent on weather and climate. Failure of rains and occurrence of natural disasters such as droughts, cyclones, floods etc., could lead to crop failure, food insecurity, famine, loss of property and life, mass migration and decline in national economy. The growing concern with possible impacts of changes in climate and weather on agriculture demands new dimensions of information by agro meteorologists. At the same time, the need for reorienting and recasting meteorological information, fine tuning of climate analysis and presentation in forms suitable for agricultural decision making and helping farmers to cope with the adverse impact of natural disasters and extreme events has become more pressing.

In addition, agriculture is one of the most important sectors heavily impacted by the natural disasters. The challenge in front of agro meteorologists around the world is, more than ever before, to more effectively integrate and deploy the skills to use climate information and products successfully in natural disaster preparedness strategies. There is also a need to develop locally adaptable agro meteorological strategies to reduce the effect of natural disasters especially in vulnerable areas, where food production is most sensitive and vulnerable to climate fluctuations.

I am happy to note that Dr. V. Radha Krishna Murthy wrote "Principles and Practices of Agricultural Disaster Management" which provides an insight into these issues for the benefit of all stakeholders of agriculture. It has several useful and easily understandable chapters through which weather based agricultural production can be sustained. The rigor and credibility of this book owes much to the unique nature of its presentation. He diligently merged many small chapters for better comprehension and shaped his works and techniques into sentences. The wide spectrum of issues reflected in the book are like the colors of rainbow, each one different from the other, but all emanating from one light. Dr. Murthy is an independent and practical, yet a real world agro meteorologist looking for more meaning in his works to serve the farmers. The depth of content in each chapter speaks much about the scientist in him and what he wants to do in future for the benefit of farmers.

I congratulate him for his efforts in bringing out this valuable and useful publication. I am sure that the book would be useful for the teachers, scientists, extension personnel, policy makers and also Indian farming community. Above all, this publication could certainly serve as a reference to students in the field of agro meteorology. I wish Dr. Murthy all the best in all his endeavors.

- Alluri Padmaraju

Preface

Yendaro mahanubhavulu, andariki vandanamulu

My parents were farmers. I as well. My mother often felt humiliated because of the continued losses that we used to incur in our agriculture. She must have lived many more years had she not worked tirelessly in our crop fields, against odd weather and climatic conditions.

In any nation people "live" on the food produced by the farmers. However, a humble and sincere request-hold your breath and listen-such farmers are ending their "lives" (committing suicides) in some states of India in general and Andhra Pradesh in particular. Humanity cannot tolerate this catastrophe. With full confidence I can state that one of the main reasons for this unfortunate situation is the overall variability of food production due to weather and climate related risks and uncertainties and subsequent un-remunerative prices.

Farming is deeply interconnected with weather and climate. Agriculture constitutes the principal livelihood of 70% of the world's population. As the silent hunger crisis has already reached a historic high with 1.02 billion people going hungry every day, raising food production to meet needs of the world population of 9.1 billion people in 2050 in light of impact of climate change may be one of the biggest challenges of the century.

I am not a philosopher. I am only an operational agro meteorologist. I have spent all my life learning agrometeorology. What I learnt in 32 years of service are as follows:

1. Weather is a non-monetary input in all agricultural operations. If weather based farming is done, the cost of cultivation of crops can be reduced at least by 10% and quality of the agricultural produce be improved by 2-3 %

2. Past 10 days weather is as important as 10 day forecast of weather, in developing weather based technologies and making farm management decisions

3. Growing Degree Days are very much useful in predicting pests and Helio-Thermal Units are highly useful in predicting diseases on crops.

I believed that when the farmers and all the stakeholders involved in crop production are enlightened on the above aspects then they would continue to produce crops profitably. Therefore, this book is written by taking relevant content and information from the works that I did in my university and CAgM of WMO, Geneva, Switzerland.

I will not be presumptuous enough that my above works can be role models to save the lives of farmers. However, if, my above mentioned findings save the life of a single farmer living in an obscure place in an underprivileged social setting, then my being is worth it. This is my ambition and that farming shall be profitable and farmers shall "live" comfortably with highest dignity. I go on following this dream.

Yendaro mahanubhavulu, andariki vandanamulu, mee anadari ashirvachnamulu naaku kaavali.

-Author

Acknowledgement

I prostrate with the highest devotion before the majestic, gorgeous, rapturous and resplendent deity, The Lord of Seven Hills, *kaliyuga daivam* Sri Tirumala Tirupati Srivenkateswara Swamy for HIS copious blessings on me in accomplishing this work.

My parents taught me that the very quality of my life is based upon how thankful I am towards GOD. I am grateful to them for their eternal blessings through noble souls and large hearts and the divine strength which enabled me to realize the task of writing this book. A high debt of gratitude to both my in-laws for their wonderful love and affection towards me. I take this opportunity to thank all my family members for their uniform affection to me.

This book could not have been written without the assistance of my wife who typed the manuscript, helped in editing with enthusiasm and patience. I thank her from the bottom of my heart. Special thanks also to her for the support and help to keep my heart open in the midst of discharging innumerable responsibilities. A quite thank to my daughter Anasuya Pooja and son Srivenkatesh, who are most unselfish and whose world is me and my wife. Their innocent faces are a treasure of strength to me.

I would like to express my immense and respectful gratitude with folded hands to Dr. M.V.K. Siva Kumar garu, formerly The Chief, Agricultural Meteorology Division, World Meteorological Organisation, Geneva, Switzerland who has been the strongest pillar of strength all through my career and personal life.

I am highly indebted to Agricultural Meteorological Division, WMO, Geneva for providing me excellent overseas opportunities to learn on operational agricultural meteorology and my university for permitting me to attend these scientific events.

I am grateful to Dr. Alluri Padmaraju garu, Hon'ble Vice- Chancellor, ANGRAU for the encouragement given to me to settle in my professional career and for writing the foreword.

-Author

Contents

Chapter 3 : Weather – Disasters – Management

Chapter 4 : Weather – Erosion – Remote Sensing – Crop Growth Models

Chapter 5: Climate Change – Agriculture

Chapter 6 : Weather Health – Crops – Farmers

Weather – Agriculture – History

Unity in variety is the plan of universe. If it be true that God is the centre of all religions and that each of us is moving towards him along one of these radii, then it is certain that all of us must reach that centre, where all the radii meet, all our differences will cease.

- Isopanishad

1.1 Introduction

Human being (*Homo sapiens*) has been on the earth for approximately 2 million years. He has been a hunter gatherer for 99.5 percent of existence and this period is considered as the most successful. Only, 12,000 years ago he started domesticating plants and recognized weather as the most precious natural resource. He managed environments in which he lived for generations by following environmental friendly agricultural practices and without significantly damaging local ecologies. This indicates that Indigenous Technical Weather Knowledge (ITWK) has immense potential to manage the risks of climate change. However, in the past 200 years of scientific agriculture there has been an over exploitation of natural resources and the environment has degraded. Therefore, it could be deduced that even in the present day scientific era alternative ways of managing resources could be made by blending this knowledge with modern techniques of managing risks and uncertainties.

History unveils that the genesis of agriculture in Asia as a means of sustaining human life can be traced back to 10,000 BC. However, in the absence of written records about the beginning of agriculture in the prehistoric Asia, one has to depend on archeo-botanical materials obtained during several archaeological excavations conducted in Asia. According to a study, 1700 BC to 1700 AD had been very important for Indian agriculture. It was observed that the Almanacs (Panchanga) have been extensively used for rainfall forecasting and the hand written panchanga of early 17th century and later periods are available in some Asian libraries.

1.2   History of Ancient Indian Agriculture with Reference to Agricultural Meteorology

- **From 10000–7500 BC:** During droughts the tribes of Central India domesticated 165 edible plants, ate seeds of 31 plants and 19 tuber crops while using honey for sugar

- **From 7500–3000 BC:** The Vedic Indo Aryan scholars composed Rig Veda (3700 BC) in which it was mentioned in 3 hymns on rain God that strong monsoon winds blows over India. Also, several observations on rainfall were recorded. It was also mentioned that 'Sun' is purifier and protector of everything on earth and provides 'Light' by which this world feed all living organisms. "Sun" in the form of fire has divided seasons characterized by heat, rain and cold. In summer, water goes up and in rainy season it comes down

- **3000–100 BC:** During this period domestication of plants and animals gave "Food Security". During 3000–1700 BC in Western India a pre-historic large mud-embankment was constructed on "Ghod" river to store flood water. Atharvana Veda (2000 BC) and Ramayana (2000 BC) mentioned about rainfall and its prediction in several versus. In 300–400 BC, "Krishi – Parashara" a very ancient and profound treatise in Sanskrit on agriculture was composed by sage Parashara which threw new insights into ancient Indian knowledge of astrology based prediction of rainfall, describing the concept of clouds and rainfall. He took the help of ruling and minister planets to estimate the sufficiency of rains in a year. He concentrated on visible causes of rainfall i.e., clouds and described four different types of clouds which differ from each other by the type of rain shed by them. The type of clouds was identified first and the amount of rainfall that the particular type of cloud would shed during that year was estimated. Susrutha (400 BC) wrote that proper season helps the emergence of strong and non-diseased sprouts. In Kautilyas Arth-Shastra (321–296 BC) there was a mention about the distribution of rainfall in different parts of India.

He also suggested that seed grains shall be exposed to night mist and day heat for seven consecutive days for better germination

- **100 BC–600 AD:** During 100 BC to 200 AD several methods of crop cultivation based on rainfall and seasons were revealed by several proverbs, village songs etc. Six seasons were mentioned in several books as early spring, late spring, cloudy, rainy, early winter and late winter. A famous Chola king Karikala (180 AD) constructed a 160 km embankment along the Cauvery river to protect his kingdom from floods. Varahamihira (505–587 AD) the first full fledged meteorologist in the history of India worked on forecasting rainfall through modeling. He wrote "Brihat Samhita". Based on lunar mansion and zodiac sign, he developed the first ever model for forecasting seasonal rainfall

- **600–1300 AD:** Kashyapa (800–900 AD) in his "Kashyapa Krishisukthi" advised that well ripened seeds shall be preserved after sun drying in heaps of straw, vessels etc., to avoid damage from water, wind and rain. Surapala wrote "Vrikshayurveda" (900–1000 AD) in which he estimated total rainfall during the rainy season. In 1120 AD Somaskhara Deva observed that seeds of a fruit ripened naturally be dried in sun for safe preservation. Certain micro meteorological weather modifications followed during this period include: preservation of winter vegetables in pits lined with wheat straw on all four sides, bottom and top and covered with soil for safer use in summer; earthen vessels painted with castor oil (acts as a barrier to moisture ingress) were used for grain storage

- **1300–1750 AD:** Mohammad Bin Tughlaq (1325–1351 AD) undertook irrigation works through construction of dams to combat "Drought". The severe drought (1398–1414 AD) resulted in developing several drought combat techniques. During Moghal period (1530–1761 AD) elaborate observations on climate and natural resources were recorded, particularly Akbar (155–1605 AD) and Jahangir (1605–1625 AD) attempted good number of climate related works for betterment of Agriculture.

1.3   Rainfall Prediction in Ancient India

1.3.1 By Astrology

The science of astrology started with understanding seasons and weather in relation to movements of planets. All cultures and civilizations have developed a form of astrology among which the Indian system is "Luni-solar" based. The main activity in this system is the time reckoning and calendar computations. In terms of weather prognostications, two important aspects are followed; the onset of rains is linked to the wind direction and the total seasonal rainfall is linked to phases of moon constellations and other planets in cycles. Astronomical calculations can be made for any number of years. As the position of planets can be predetermined the rainfall is forecast for any time in future.

The principles Governing Rainfall Prediction by Astrology

Water is classified based on its site of availability and use. Each site is a living entity and certain effects are brought about by organic action. Rain is conceived by a site and the gestation period is 192 ± 2 days. The position of planets at the entry point of moon in a particular constellation decides the probability of conception of rain and chances of normal delivery (1, 2, 3 and more days).

The procedure of Rainfall Prediction by Astrology

A fixed reference place/point on the earth is drawn on a chart. The exposure of this point to the sun during the day, the periodic waxing and leaning of moon, the regular appearance and disappearance of planets are projected as celestial cycle on this chart. The planetary chart laid out for the time when the sun influences (enters) each aesterium is noted and the angle between two planets and most powerful position in different angles is computed. The primary planetary chart of time and date of Chaitra Pratipada, Ashwini is configured. Actual predictions are derived using the house of each planet and its angle with reference to other planets. More so, the planets which are more powerful during the year with special reference to rainfall will be watched carefully.

The unique feature of Indian astrology is "Capsular theory". The two asterium/lunar mansions are arranged in capsules of 2, 3 or 7 and their associated effect on weather is predicted. A sequence is followed in serpentine fashion and more planets in any capsule gives the effect of capsule. The Neptune, Moon and Venus are universally accepted as responsible for rain.

1.3.2 By Panchanga

In India, the classical Hindu almanac is known as "Panchanga". It is a book or record of astronomical phenomena containing a calendar of days, weeks and months of the year. Weather prognostications and seasonal suggestions for a state or country are often mentioned. It acts as an astronomical guide to farmers to start any farming activity. Panchanga making might be traced from Vedic literature more so during Vedang Jyothish period (1400–1300 BC). However, hand written panchanga of early 17th century and later periods are available in some Indian libraries.

The word "Panchanga" is derived from Sanskrit language in which "Panch" means 'five' and 'ang' means "body parts". There are several ways through which rainfall is predicted by Panchanga. Of them a few prominent are lunar day (30 days in one month);   week (7 days); aesterium constellation (Twenty seven days);   time (The twenty seven number of times);   Joint motion of sun and moon covers the space of aesterium etc. The permanent relationship is established among all the five parts and printed in the form of a guide every year. This is useful to practice agriculture and other weather related activities by the farmers. In general, the amount of rainfall in the coming months/seasons/year is assessed on the basis of symptoms at cloud conception, positions of the sun and moon in a particular division or zodiac and related considerations. There are several ways through which rainfall is predicted by panchanga. of them a few prominent are:

A. *Type of cloud*

Panchanga identifies the different types of clouds and based on dominance of a particular type in a year the rainfall is predicted.

S. No.	Type of cloud	If dominant in a year
1.	Abartak	Rain will be received in certain places in that year
2.	Sambartak	Rain will be received in all parts of the country
3.	Pushkara	Rainfall quantity will be less
4.	Drona	Abundant rain water

B. *According to ruling planet*

The rainfall prediction based on ruling planet is as follows; Sun (Moderate), Moon (Very heavy), Mars (Scanty), Mercury (Good), Jupiter (Very good), Venus (Good), Saturn (Very low and stormy wind).

C. *According to capsular Theories*

To predict the monsoon three different capsular theories are as follows: Bi-capsular, Tri-capsular, Seven capsular. This theory is based on grouping of all asteriums into 2, 3 and 7 categories according to specific criteria. Then based on union of different genders of planets and planetary conjunctions, future rain, immediate rain etc., are predicted.

Several astrologers have developed almanacs and predicted rainfall distribution based on which cropping patterns and area were decided in the past. These almanacs had helped the farmers in ancient times in many ways.

1.3.3 Ancient Models for Prediction/ Forecasting Seasonal Rainfall in Ancient India

A. As detailed in Brihat Samhita

Varahamihira tried to evolve a technique based on astrology in which he was proficient. His technique lays down that after the occurrence of full moon day of the month of "Jyestha" (June of Gregorian calendar), the asterium or lunar mansion or "Nakshatra" of the day on which the first rainfall of that year rainy season is received should be noted. This asterium provided the basis for the forecast of seasonal rains. There were twenty seven such asteriums or lunar mansions in Indian astrology, with each one falling under a particular zodiac sign.

B. As detailed in Krishi Parashara

Of the 243 verses in Krishi Parashara 69 relate to prediction of rainfall have strong astrological content. For the prediction of rainfall in the whole year Parashara has given methods based on the "Ruling Planet" and the "Minster Planet" of the year; the type of cloud, the direction of wind, the change in level of river water on a specific day and star constellation. Each model he developed was simple and farmer with basic knowledge of calendar could learn it easily by memorizing the verses.

1.4 Indigenous Technical Weather Knowledge (ITWK)

1.4.1 Evaluation of ITWK on Rainfall Forecasting

The term "Indigenous Technical Weather Knowledge (ITWK)" is used as synonym to 'local' and 'traditional' knowledge to differentiate it from 'scientific', 'modern' and 'rational' knowledge. The ITWK products of weather in Asia are strong knowledge pools developed by different communities through keen observations, natural selection and centuries of trial and error. Those ITWKs that were proved effective and sustainable on a long term basis to counter the extreme weather events and helped reap better harvest existed through hundreds of years of adaptive evolution in Asia. Even though in many cases the ITWK on weather is 'unwritten' it did exist in different brains, languages and skills with innumerable communities and cultures because the ITWK helped for successful and persistent adoption to the environment.

1.4.2 Examples of Agrometeorological Services using ITWK

- Rainfed rice cultivation in the eastern state of West Bengal, India is highly complex, risky and uncertain due to insufficient or excess early or late rains, early or late floods, the erratic withdrawal of monsoon etc. Therefore, the farmers adopt certain traditional methods to minimize the losses of crops due to these weather abnormalities. They sow a mixture of both autumn and winter rice varieties to ensure that a good harvest of at least one variety in the event of these weather aberrations. Also, they use staggered nursery for transplanting in the main field because the cost of nursery rising is less and under flooded conditions they also transplant lengthy and aged seedlings to overcome the submergence. Similarly, to meet the optimum date of sowing they sow the seed as soon as the summer rains are received and take the advantage of initial soil moisture for germination

- In dryland areas of northern India the delayed, erratic distribution and untimely rains are the serious problems for cultivation of wheat in winter (November) and maize in rainy season (June). In winter, if excessive early rains are received it will not be possible to sow wheat crop because, due to low air and soil temperatures the drying of excessive soil moisture takes atleast one month. Therefore, traditionally farmers follow "dry sowing" method in which the seeds are sown in the soil by taking the advantage of initial soil moisture of late SW monsoon rains. The seeds remain healthy in the soil and germinate when enough rains are received. Similar method of early sowing in May is followed for maize crop also, which helps the farmers to overcome the difficulty of sowing number of times and meeting the optimum date of sowing in the event of untimely rains

- Cultivation of cucurbits, melons, gourds etc., in sandy soils of southern, northern and north western India is risky and uncertain. In northern India the seedlings of these vegetables are grown in poly bags and allowed to grow in trenches to protect them from frost damage. These seedlings are transplanted in the main field during February second fortnight and farmers reap bumper harvest. In contrast, in southern India the fully grown fruits of these vegetables are put in trenches along with their tendrils and twigs to prevent their cracking and damage due to high air and soil temperatures

- The Indian dryland farmers also developed another ITWK to overcome the menace of heavy infestation of pests and diseases on vegetables grown in off-season. In northern India the farmers raise nurseries of vegetables in November (early winter) and take the advantage of enough soil temperature for germination. However, the growth of these seedlings is restricted till January due to low soil and air temperatures in winter. The transplanting of seedlings are taken up in February such that the main field crops grow under optimum weather conditions and yield potentially. If the seedlings are grown in February (late winter) then the flowering and fruiting synchronizes with the onset of monsoon rains and the pests and disease menace cause heavy or total loss of crops.

The above methods were being followed in India since times immemorial.

1.5   India Meteorological Department

(Adapted from "Agrometeorological Services in India", Pune. A publication of "Agricultural Meteorology Division", IMD, GOI, MOES, Pune.)

Although, India Meteorological Department (IMD) was established in 1875 subsequent to a disastrous tropical cyclone hit Calcutta in 1864 and the famines in1866 and 1871 due to the failure of the monsoons, the rainfall data/observations were started well before. India is fortunate to have some of the oldest meteorological observatories of the world, that include Madras (now known as Chennai), established in 1793, Bombay (now known as Mumbai) in 1823 and Shimla in 1841. With the gradual growth in the expansion of observational network varieties of data have been generated and accumulated in a span of many years. The daily rainfall data of all the rain gauge stations of India i.e., Mumbai from

1847, Bandra from 1855, Deesa from 1856 etc are available at IMD. The country is divided into 36 meteorological sub-divisions and long series of monthly, seasonal and annual rainfall data from 1875 onwards are available.

First long range forecast of monsoon was initiated by Sir H.F. Blanford in 1886. IMD has been issuing operational monsoon forecast regularly for the country since 1988 and subsequently started four homogenous regions.

1.5.1 Agrometeorological Services in India

Weather and Agriculture

Indian agriculture has, for centuries, been dependant on the weather and the vagaries of the monsoon in particular. Uncertainties of weather and climate pose a major threat to food security of the country. Extreme weather events like heavy rains, cyclone, hail storm, dry spells, drought, heat wave, cold wave and frost causes considerable loss in crop production every year. An efficient use of available climatic resources, besides soil and water resources, minimizes the adverse effect of extreme weather and makes benefit of favourable weather. Weather services provide a very special kind of inputs to the farmer as advisories that can make a tremendous difference to the agriculture production by taking the advantage of benevolent weather and minimize the adverse impact of malevolent weather.

Weather Services to Agriculture

In order to provide direct services to the farming community of the country an exclusive Division of Agricultural Meteorology was set up in 1932 under the umbrella of India Meteorological Department (IMD) at Pune with the objective to minimize the impact of adverse weather on crops and to make use of favourable weather to boost agricultural production.

The major activities of the Division are:

- Technical Assistance
- Research and Development
- Services
- Human Resource Development.

Integrated Agromet Advisory Services

To meet the farmer's need in real-time and to have a state-of-art Agromet Advisory Service (AAS), Integrated Agromet Advisory Service in the country involving all the concerned organizations viz., Indian Council of Agricultural Research (ICAR), Ministry of Agriculture (Centre and State), State Agricultural Universities (SAUs) and other agencies has been started from April 2007.

Collaborating Agencies under IAAS

State Agricultural Universities (SAU)

Indian Council of Agricultural Research (ICAR) and its research institutes

Indian Institutes of Technology

State and Union Departments of Agriculture

Prasar Bharati and other media (Radio, TV and Print).

Network of Observatories

The Division of Agricultural Meteorology maintains and provide technical support to a wide range of agromet observatories from where different kinds of data on agromet paramaters are generated.

AAS *Bulletins are issued from three levels*

- National Levels by National Agromet Advisory Service Centre, Agrimet Division, IMD, Pune
- State Levels by State Agromet Service Centre at Regional Meteorological Centre/Meteorological Centre (23)
- District Level by Agromet Field Units (130).

District level agromet advisory bulletins are issued for the farmers. The State level composite AAS Bulletins are issued for State level planners e.g. State Crop Weather Watch Group (CWWG) and other users like fertiliser industry, pesticide industry, irrigation department, seed corporation, transport and other organizations which provide inputs in agriculture. The National Agromet Advisory Bulletins are primarily targeted for national level planners e.g. CWWG, Department of Agriculture and Cooperation, Ministry of Agriculture, New Delhi and also communicated to all the related Ministries (State and Central), Organizations and NGOs for their use.

Main features of AAS *bulletin*

- Significant past weather
- Quantitative weather forecast for next five days
- Advisories for farming community.

Broad Spectrum of Agromet Advisories

- Sowing/transplanting of Kharif crops based on onset of monsoon
- Sowing of rabi crops using residual soil moisture
- Fertilizer application based on intensity of rain
- Delay in fertilizer application based on intensity of rain
- Prediction of occurrence of pest and disease based on weather
- Propylactive measures at appropriate time to eradicate pest and diseases
- Weeding/thinning at regular intervals for better growth and development of crops
- Irrigation at critical stage of the crops
- Quantum and timing of irrigation using meteorological threshold
- Advisories for timely harvest of crops.

IMD has started issuing quantitative district level (612 districts) weather forecast up to 5 days from 1st June, 2008. The products comprise of quantitative forecasts for 7 weather parameters viz., rainfall, maximum and minimum temperatures, wind speed and direction, relative humidity and cloudiness. In addition, weekly cumulative rainfall forecast is also provided. IMD, New Delhi generates these products based on a Multi Model Ensemble technique using forecast products available from number of models of India and other countries. These include: T-254 model of NCMRWF, T-799 model of European Centre for Medium Range Weather Forecasting (ECMWF); United Kingdom Met Office (UKMO), National Centre for Environmental Prediction (NCEP), USA and Japan Meteorological Agency (JMA).

The products are disseminated to Regional Meteorological Centres and Meteorological Centres of IMD located in different States. These offices undertake value addition to the products and communicate to 130 AgroMet Field Units (AMFUs).

The district level agromet advisory service is multi-disciplinary and multi-institutional project. For this, IMD organises District Agromet Advisory Service meeting in each of the state of the country inviting the officers/scientists and all the stakeholders with objectives to create appropriate information generation-cum-dissemination mechanism as well as extension mechanism at district level for communicating the agromet advisories to the farmers regularly.

Dissemination of Advisories

Dissemination of Agromet advisories is done through

- All India Radio (AIR) and Doordarshan
- Private TV and radio channels
- Newspapers
- Internet
- ICAR and other related institutes/Agriculture Universities/Extension network of State/Central Agriculture Departments
- Krishi Vigyan Kendras
- Advisories are delivered to the end users without any delay
- Interactive tuning of advisories with the farmers/managers as frequently as possible
- Disseminated in English and local languages/dialects and is easily understandable by farmers.

Linkages between Districts Agriculture Offices (DAOs) and AMFUs level are being developed for effective dissemination of advisories at district, block and village levels.

Extension of Advisories

Extension wing of the State Departments of Agriculture, State Agricultural Universities, Indian Council of Agricultural Research Institutes are working for application of the advisories in the farmer's fields.

Feedback and Awareness of Agromet Service

In order to improve the quality of the agromet advisory services, regular direct interactions are being made by the AMFUs with the farmers, State AAS units and Agrimet Divisions are regularly participating in Kisan Melas, farmer's gatherings etc., to interact with the farmers personally and collect the feedback from farmers. Roving seminars are being organized in different States by AMFUs to create awareness about usefulness of weather/climate information, agromet advisory services among the farming community.

Crop Yield Forecasting

A need for quantitative crop yield forecast outlooks has been felt for quite sometime. A beginning towards its realization has been made by undertaking a study of past crop yield in relation to meteorological parameters, principally rainfall and temperature. Based on these studies quantitative crop yield forecast formulae have been developed for 22 sub divisions in the country for Kharif rice and 9 sub divisions for wheat. The tentative forecast for crop yield is being issued every month during the crop season using this methodology.

Research and Development

From the inception the Agrimet Division is working on research and development programmes to strengthen the operational agro-meteorological services in the country.

Microclimatic Studies

The study of microclimate of crops was one of the earliest investigations under taken by the Agricultural Meteorology Division in Pune since 1932. The subject of micro-climates received intensive attention and significant contributions have been made in this field of research. Research activity in the field of microclimatology was carried out extensively by various personnel from IMD as well as others. Systematic observations on the characteristic micro-climates of the air layers close to the ground in the open and inside various crops have been recorded and a large volume of micro-climatological data were collected.

Agroclimatic Classification

In order to bring out agricultural potential of a region, its agroclimatic classification has to be made. Considerable work has been done on agro-climatic classification. Agroclimatic zones have been delineated using Thornthwite moisture availability index and other methods. Penman method has been used to estimate evapotranspiration of about 230 stations located in India. These estimates are used to compute water balance of these stations in India and ultimately agroclimatic classification.

The Agricultural Meteorology Division has made a detailed examination and prepared suitable diagrams of the frequencies of occurrence of various adverse weather phenomena (like hail storms, frosts etc.) that affect growing crops, extremes of temperature met with in summer and winter, estimated evaporating power of the atmosphere etc. Such diagrams help to show how often the farmer may be called upon to mitigate the effects of adverse weather phenomena by resorting to possible protective measures like artificial heating, use of wind brakes etc.

Cropping Patterns

By analyzing the rainfall records of 2000 stations for 70 years, the periods and amounts of "assured rainfall" have been worked out for various regions particularly in the dry farming tract of India. The length of dry spells and wet spells during the monsoon, drought proneness and agro-climatic classifications have also been studied with climatological data. This information is helpful in choosing appropriate crops for various regions, determining the most favourable growing seasons for rainfed crops and selecting drought tolerant crop strains.

Sowing dates

Optimum dates for sowing have been determined for the States of Maharashtra, Rajasthan, Gujarat and Madhya Pradesh, by using daily rainfall data from 1901 onwards. Such information helps in deciding the best period for sowing operations, water conservation measures and evolution of appropriate cropping patterns.

Even in those parts of the country where irrigation facilities does not exist, the crop production can be maximized by better scheduling of irrigation. Scarce resources can be economically used by providing water to crops when it is known to be most beneficial. For this, precise water requirements of crops at various growth stages are being studied through field experiments and regular lysimeter measurements of evapotranspiration.

Remote Sensing Applications

Satellite remote sensing techniques have been used for acreage estimation of jowar. Spectral response of crops at various growth stages and states are studied to help crop identification. Soil moisture studies using micro-wave remote sensing technique were made.

Crop Weather Analysis

Theoretical models of crop weather relationship enable to understand, quantitatively, the role played by weather elements on crop growth and yield. Variability of soil moisture and soil temperature and the contribution of dew have been studied in relation to crop growth. Fluctuations in weather with regard to crop factors like leaf area index, stomatal resistance, crop coefficient and dry matter production have been studied. Energy balance of the crop canopy for cereals and legumes are being worked out. A number of crop weather calendars were prepared based on crop weather studies. Crop growth simulation models (DSSAT) are being used to develop crop weather relation as well as crop yield forecasting.

Considerable research has been done on the weather conditions conducive to outbreaks of crop pests and diseases like paddy stem borer, jowar shoot fly, cotton bollworm, sugarcane borers, groundnut tikka, potato beetle and wheat rusts. The results help to organize timely crop protection measures with optimum use of expensive chemicals. The desert locust breeding and invasion has been extensively studied in relation to soil and weather conditions.

Drought Studies

By analyzing rainfall data since 1875, the probability of occurrence of drought in various parts of India has been worked out. Different parameters like water availability, soil moisture stress, aridity index have been studied. Droughts are monitored by deriving aridity anomalies on a fortnightly basis in the Kharif season and weekly basis in the northeast monsoon season over the southern peninsula.

Dry Land Farming

The area having annual rainfall between 40-100 cm and practically with no irrigation facilities is known as dry farming tract. Dry farming tract comprise 87 districts and is spread over Haryana, Punjab, Rajasthan, Gujarat, Uttar Pradesh, Madhya Pradesh, Maharashtra, Andhra Pradesh, Karnataka and Tamilnadu. Extensive research was carried out on assessment of short period rainfall probability, compilation of frequency, duration and intensity of dry and wet spells, assessment and seasonal and diurnal variation of meteorological parameters, derivation of agroclimatic zones and sub-zones.

Weather and Phenology of Crops

Phenology is the science which deals with the recurrence of the important phases of animal and vegetable life in relation to the march of seasons during the year. The dates of manifestation of phenophase constitute an integral of climatic effects as they take into account the weather over past periods and also the weather at the moment. Studies were made to observe the effect of climatic factors on the flowering, fruiting and maturity of four trees i.e., manblack, neem, tamarind and babul. The observations were taken at about 200 phenological stations located in the agricultural farms, soil conservation centre and meteorological stations.

Agrometerorological observations with State of Art Instruments

A number of experiments using portable photosynthesis system, infra-red thermometer, portable leaf area meter, dew point generator and dew point microvoltmeter were conducted at College of Agriculture, Pune during rabi season on impact of CO_2, photosynthetically active radiation (PAR), temperature and relative humidity on the rate of photosynthesis and other parameters like transpiration, stomatal conductance etc., in field crops like maize, jowar, safflower, sunflower and soybean.

Training Programmes in the Division

Following training programmes for national and international training for agromet observers, university teachers/departmental officers and foreign personnel are regularly arranged in the Division.

- Foreign Trainees' course of 6 months duration
- Agromet core course of 3 weeks duration for teachers and scientists of agricultural universities/institutes
- Observers course of 3 weeks duration for recording observation
- Departmental trainees (Grade 'A' officers)
- On the job training for advanced (revised) meteorological course trainees
- Basic agromet course of 3 weeks for Departmental candidates
- Refresher course of 2-3 weeks for departmental/non-departmental officers.

Future Programme

Proposed modes of dissemination – Dissemination through Common Service Centres (CSC) by DIT

- Department of information Technology (DIT) is planning to develop ICT facilities for the benefit of the citizens, especially those in rural and remote areas. 1,00,000 CSCs will be set up at village level shortly to provide all possible services. Among others, agromet services will also be provided through the CSCs
- Ministry of Agriculture is already operating "Agricultural Technology Management Agency (ATMA)" project in several districts on the line of NGO concept
- It is also planned to provide AAS link to Village Knowledge Centers at taluka level that have already been opened by M.S. Swaminathan Research Foundation and Alliance for providing the need based information at village level

- Under Integrated Agromet Advisory Service (IAAS) Scheme, IMD is exploring to tie up with different public and private organizations to use IVR and SMS technology which are already working in dissemination of agricultural information to the rural villages.

1.6 World Meteorological Organisation

(Adapted from "The World Meteorological Organisation at a Glance". A WMO Publication, Geneva, Switzerland)

1.6.1 Scope of World Meteorological Organisation

Weather, climate and water know no political boundaries. To promote international cooperation in these areas, the World Meteorological Organisation (WMO) coordinates the activities of the National Meteorological and Hydrological Services (NMHSs) of its 188 Members. Originating from the International Meteorological Organisation established in 1873, WMO was created in 1950 as an intergovernmental organization and became a specialized agency of the United Nations in 1951.

- The National Meteorological and Hydrological Services work around the clock, all year round, to protect and provide vital information to communities
- The early and reliable warnings of the occurrence of severe weather, air quality and climate events allow decision makers, communities and individuals to be better prepared, which help save life and property, protect resources and the environment and support socio-economic growth
- WMO ensure that all nations are able to take full advantages of weather, climate and water information and products for their sustainable development and the safety and well-being of their people.

1.6.2 Functions of WMO

Improving Safety and Well-being

- The WMO galvanizes the global community to improve understanding of weather, climate and water
- The WMO provides unique mechanisms for the timely exchange of data, information and products

- The WMO makes major contributions to sustainable development, the reduction of loss of life and property caused by natural hazards related to weather, climate and water, as well as safeguards the environment and the global climate for present and future generations

- The WMO through its members provides forecasts and early warnings to nations, economic sectors and individuals, that help prevent and mitigate disasters, save lives and reduce damage to property and to the environment through better risk management

- The WMO draws world attention to the depletion of the ozone layer, climate variability and change and their impacts, dwindling water resources and air and water quality

- The WMO monitors and forecasts the transport of chemical and oil spills, forest fires, volcanic ash, haze and nuclear isotopes. It assists in the formulation of global and regional strategies, conventions and the implementation of related action plans.

Taking the Pulse of the Earth System

- WMO provides up-to-date, accurate and quantitative information on the state of the Earth's atmospheric system, the oceans, surface water bodies and underground water. It also monitors the interaction of the atmosphere with the Earth's surface, ecosystems and human activities

- WMO facilitates the provision and exchange of near-real-time information from across the globe around the clock. The data are collected by 10000 land stations, 3000 aircrafts, 1000 upper-air stations and more than 1000 ships working in tandem with 188 National Meteorological Centres and 35 Regional Specialized Meteorological Centres. These are bolstered by over 16 meteorological, environmental and operational satellites and 50 research satellites

- The WMO Integrated Global Observing Systems (WIGOS) acts as an umbrella for these observational networks, using the WMO Information System (WIS) to connect together all regions for data exchange, management and processing

- The WMO research programme coordinate and integrate the research activities of Members to take full advantage of global observations in analysis of the weather and climate and to develop computer models that represent the key underlying processes for improving the accuracy and range of weather forecasts

- WMO supports air quality services, measurements of hydrological variables and ensures that observational and monitoring instruments everywhere are accurate and provide standardized data generated in one place are to be usable elsewhere in the world

- WMO also assists countries in enhancing their data-management capacity. Data-rescue activities help NMHSs, especially those of developing countries, access historical data for various purposes

- WMO monitoring observation systems are a core contribution to the Global Earth Observation System of Systems (GEOSS), aimed at developing a comprehensive, coordinated and sustained international approach to understand and address global environmental and socio-economic challenges.

Research

- WMO coordinates and organizes research programmes that contribute to scientific understanding of the dynamical, physical and chemical processes in the atmosphere and oceans, as well as the interactions of various components of the Earth system on all time and space scales

- WMO promotes research into fundamental scientific understanding of the physical climate system and climate processes needed to determine to what extent climate can be predicated and the extent to which humankind influences climate

- It promotes the advancement of atmospheric sciences in understanding atmospheric compositing changes and consequent effects on weather, climate, urban environment and marine and terrestrial ecosystems. The WMO Atmospheric Research and Environment Programme accelerates improvements in "nowcasting" – forecasting the next six hours – and one day to two week high impact weather forecasts for the benefit of society, the economy and the environment

- It also focuses on tropical cyclones and monsoons. Other programmes aim to measure and understand the influence of greenhouse gases and other climate changing particles and chemicals in the atmosphere

- Climate research on global to regional scales and time horizons ranging from weeks to centuries is coordinated by the World

Climate Research Programme (WCRP) co-sponsored by WMO, the International Council for Science and the Intergovernmental Oceanographic Commission of UNESCO

- WMO has been one of the leaders of the International Polar Year (2007-08) and helps Earth's polar regions to enable a better understanding of our future climate, among other things.

1.6.3 Applications

Weather, climate and water impact many socio-economic sectors, agriculture and fisheries, energy, transport, health, insurance, sports and tourism.   WMO's endeavours to promote the application of meteorological, climatological, hydrological and oceanographic information to human activities are therefore of great importance worldwide.

Disaster Prevention and Mitigation

About 90 per cent of all disasters are related to weather, climate or water. The human and material losses caused by natural disasters are a major obstacle to sustainable development and world safety and security. With other international, regional and national organizations, WMO coordinates the efforts of NMHSs to improve forecast services and early warnings to protect life and property from natural hazards, such as tropical cyclones, storms, floods, droughts, heat and cold waves and wildfires.   In addition to public safety, such extremes affect water and food supplies, the environment, transport and many other socio-economic sectors.

Emphasis is on improved warnings and better integration of such information in disaster risk management: one dollar invested in better prediction and disaster preparedness can prevent seven dollars' worth of disaster related economic losses; a considerable return on investment. WMO's objective is to reduce by 50 per cent, by 2019, the associated 10-year average fatality of the period 1994-2003 for weather, climate and water related natural disasters.

Water Resources Assessment and Management

Global freshwater resources are both diminishing and deteriorating under demographic and climate pressures. Water is essential for life, for generating hydroelectric power and meeting irrigation and domestic requirements. WMO promotes water resources assessment and provides the forecasts needed to plan water storage, agricultural activities and urban development. It supports an integrated, multidisciplinary approach to managing water resources.

Agriculture and Food Security

The agricultural sector critically depends on timely and accurate weather, climate and water information, particularly as it faces increasing climate risks. The observations, analyses and forecasts produced by WMO Members enable the agricultural community to increase crop and livestock yields, plan their planting and harvest time, and reduce pests and diseases. Regular Regional Climate Outlook Forums, as well as training, coordination services and resources provide a range of services to improve agricultural output and sustainability and contribute to world food security.

Public Health

Through its Members, WMO provides weather and climate services to the public health community. Early warnings for disease epidemics, disaster prevention and mitigation and air quality services all aim to protect people's health and welfare. Several Regional Climate Outlook Forums, for example, now support malaria surveillance and warning systems in Africa. Heat health advisory services give early warning of heat waves. Joint partnerships with international, regional, and national health sector partners are increasing the effective use of weather and climate information in support of such efforts.

Transport

The aviation sector requires a range of information on weather conditions. Precipitation, wind, turbulence, fog and a host of other factors affect day-to-day aviation activities. WMO ensures the worldwide provision of cost effective and responsive meteorological services in support of safe, regular and efficient aviation operations. Likewise, WMO provides services in support of the safety of marine and land transport. These services provide early warning to the offshore oil and natural gas infrastructure, there by aiding energy security and transport.

Oceans

WMO promotes the protection of the marine environment and the efficient management of marine resources, based on the timely collection and distribution of marine meteorological and oceanographic data. WMO provides assistance to Members in establishing national and regionally coordinated systems to ensure that the loss of life and damage caused by tropical cyclones are reduced to a minimum. It also supports the sustainable operation of fisheries through weather and climate observations and analyses.

Energy

Climate, weather and water information supports optimal development and use of renewable energy resources such as hydropower, wind, solar and biological energy. Such information also underpins the routine operation of nuclear power plants, coal power plants and other forms of energy production. WMO facilitates the exchange of data that can help energy developers and managers better plant for changes in energy demand, the development of local energy systems and compliance with environmental requirements.

Socio-Economic Development

Through its various activities, WMO helps developing countries manage resources, prevent disaster and adjust to climate variability and change. To address the special problems and needs of Least Developed Countries (LDCs), the Fourteenth World Meteorological Congress established the WMO programme for LDCs in May 2003 to enhance the capacity of NMHSs to contribute effectively to the socio-economic development of these countries. In line with the overall Programme of Action for the LDCs for the decade 2001-2010, adopted by the third United Nations Conference on the LDCs, the WMO Programme for the LDCs encompasses the five following strategic areas: fostering a people-centred policy framework; strengthening productive capacities; building human and institutional capacities; reducing vulnerability and conserving the environment; and resource mobilization.

WMO supports developing countries, and the LDCs in particular, in their social and economic development and combat against poverty by enhancing the capacities and capabilities of their NMHSs. Capacity building in the most vulnerable communities ensures a greater ability to monitor weather, climate and water conditions and plan for future conditions. One dollar of investment in weather information returns 10 dollars worth of socio-economic development. Such actions contribute to the achievement of the UN Millennium Development Goals by 2015, and especially the eradication of extreme poverty and hunger.

1.6.4 Sharing Expertise and Building Capacity

- WMO assists the NMHSs, especially those of developing countries, in their efforts to contribute, in the most effective manner, to the development plans of their countries and to become full partners in global collaborative efforts

- WMO helps its Members develop human resources through training, the provision of educational material and the awarding of fellowships. Its more than 30 Regional Meteorological Training Centres, along with a network of cooperating universities and advanced training institutions, contribute to the global effort

- WMO promotes and facilitate technology transfer, as well as the establishment and development of specialized centres of excellence in various regions.

2

Weather Elements – Measurement

Each soul is potentially divine. The goal is to manifest this divinity within by controlling nature, external and internal. Do this either by work or worship or psychic control or philosophy by one or more or all of these and be free. This is the whole of religion. Doctrines or dogmas or rituals or books or temples or forms are but secondary details.

- Isopanishad

The primary aim of agricultural meteorology is to extend and fully utilize the knowledge of atmospheric and related processes in order to optimize sustainable agricultural production with maximum use of weather resources and with minimum damage to the environment. This entails improving the quantity and quality of agricultural crops. The secondary aim concerns the conservation of natural resources. Therefore, all aspects of local weather and climate have to be measured in agrometeorology. These include standard official synoptical, climatological and those meant for special agrometeorological purposes.

2.1   Solar Radiation

Energy is the main prerequisite for life. All energy for physical and biological processes on the earth comes from sun. This is the only eternal source and life depends on it. All matter at a temperature above absolute zero imparts energy to the surrounding space. This transferring of energy and its mode of transfer is also known as "Radiation". The energy received from the sun is known as "Solar radiation". It is converted into biomass through photosynthesis on which all types of life survive. Solar radiation is important in several ways among which the following are very important for all the agricultural crops:

- Directly effects photosynthesis and vitamin synthesis
- Indirectly provides heat

24

- Powers the climate system in which crops survive
- The quantity and quality of solar radiation determines the existence and production of crops.

The total radiation flux within a given site is highly variable, changing with:

(a) The time of the day

(b) Season

(c) Weather

In addition, the variations of the total radiation flux from one site to another site on the surface of the earth are enormous. The distribution of agricultural crops responds to this variation.

2.1.1   Effects of Surface Geometry on Solar Radiation

The effects of small terrace irregularities such as mounts, ridges, furrows and trenches etc., alter the amount of solar radiation (in particular, direct radiation) striking the surface, thereby, soil temperature and soil moisture are altered. The germination and emergence of a considerable range of crops are also influenced.

Day time geometrical relationships between solar radiation and slope and aspect of small ridges and furrows depend on their orientation, with respect to slopes and position of the sun. In the early morning the tops of ridges are colder than at ground level, both because of excessive sky view and because of the (relatively) large distance to deep soil heat. Thereafter, sunny slopes heat up appreciably, depending on orientation, also because reflected radiation may still be absorbed at the other side (radiation trapping). An interesting situation occurs on very dry day. North-South oriented ridges where plants placed on west-facing slopes survive because in the early shade they make use of occasional morning dew. Nocturnal temperature relationships in small hollows, trenches (with linear and vertical dimensions of a few tens of centimetres at most, typical of furrows) may be different from those observed in cases of dimensions of tens of meters. At the smaller scale, few factors need attention:

1. The gravitational flow of cold air into shallow "frost hollows"
2. Reduction in turbulent exchange because of the shape of the hollow

3. Shortening of the period during which the sun actually can reach the soil surface (apparent day length)

4. The decrease in long-wave loss due to diminished exposure to the sky

5. The heat supply from the sides of the hollow.

Factors 1, 2 and 3 tend to cause lower temperatures, while 4 and 5 tend to reduce the nocturnal fall in temperature, however factor 5 will only work in small scales. The above techniques help to reduce the soil temperature at micro level in summer season. Therefore, vegetables such as tomatoes, chillies etc., can be grown successfully in summers.

2.1.2   Crop Management and Layout

Crops can be raised more successfully and profits could be increased by covering the soil with mulch or changing its surface shape as compared to crop raising under natural conditions.

When the crop covers the ground completely, the absolute value of temperature, humidity, winds at the surface and the vertical gradients of these elements are decisively altered by the crop itself. On horizontal ground the spacing of row crops and the seasonal changes in the ratio of crop height to distance between the rows, drastically affects the amount of sunshine reaching the actual soil surface (radiation phenomena depend upon the ratio of direct and diffuse sunshine). Crop yield per unit area is affected by both the density of planting and the production of individual plants. If densities are higher than optimal, then the light, which can still penetrate to lower parts of the crop, suffers too much intensity reduction in PAR part of the spectrum. In addition, movement of air within the crop decreases with increasing density which may reduce the CO_2 concentration near lower leaves and limit photosynthesis on sunny days with weak winds. Density of rows (spacing) may have little effect if sufficient water is available. However, in arid conditions wide spacing is recommended.

2.1.3   Measurement of Solar Radiation and Sunshine

Global solar radiation (direct and diffuse solar radiation) is measured with pyranometers containing thermocouple junctions in series as sensors. The sensors are coated black to have uniform thermal response at all spectral wavelengths. With filters, non-PAR radiation can be measured and the difference between solarimeter outputs with and without filters gives PAR data. Solid-state sensors (photo-electric solar cells, photo-emissive elements, photo resistors etc.,) may be used where radiation can be assumed to have constant spectral distribution (e.g., solar radiation within limits). Different types of *photometers and ultra-violet illuminometers*, which are adoptions of these, are used in agrometeorological research.

Light, indispensable for photosynthesis, is one of the major components of the short wave radiation. What is measured with a lux meter is not light intensity, but luminance, that is defined as "luminous flux density intercepted per unit area". Quantum sensors which measure the PAR directly in the range 0.4 to 0.7 micrometers are available. Ideally crop profile measurements with quantum sensors should be done on perfectly clear or uniformly overcast days. However, if this is not possible the problem is partially overcome by expressing the values at each level relative to the incident radiation. These profiles are compared with leaf area profiles when the light requirements of crops are being studied.

Tube radiometers for use in crops and agroforestry are inherently less accurate than instruments with a hemispherical dome, but, can be of great use in estimating the average radiation below a crop canopy or mulch relative to that above it. When mounted N – S, the sensitivity varies with the angle of the solar beam to the axis, particularly in the tropics. This adds to errors due to high ambient temperatures under low wind speeds and errors due to condensation inside the tubes. Calibrations as a function of time and ambient conditions can largely cope with such errors but filtered tubes for photosynthetically active radiation appeared unreliable in the tropics. To measure the fractional transmission of solar radiation through a crop canopy, a number of tubes are placed beneath the canopy. Their numbers and arrangement depend on the uniformity of the crop stands. A reference measure of incident solar radiation above the canopy is needed. For crop studies, the output for each tube is usually integrated over periods of a day or longer during the growing season. Integrators or loggers are ideal for this purpose. The values of fractional interception are subsequently calculated from the integrals.

Pyrgeometers are used for the measurement of long-wave radiation from the earth and surface temperature radiometers for measurements of infrared radiation emitted from near or remote surfaces. The latter are mainly used as hand held remote sensors to measure temperatures of radiating irregular surfaces such as soil, plant cover and animal skin and require knowledge of the emissivity coefficient of the observed surface.

Net all wave radiation (net flux of downward and upward total radiation i.e., solar, terrestrial and atmospheric) is measured with black coated heat flux plate sensors, in which thermocouples are embedded to measure the temperature difference between the two sides of a thin uniform plate with well-known thermal properties. Errors due to convection and plate temperature are avoided by using forced ventilation, appropriate shields and built-in temperature compensation circuits. *Net radiometers or net pyrradiometers or net exchange radiometers or balancemeters* may have a standard (about 6 cm) diameter for regular use or a miniature (about 1 cm) diameter for special work on radiation exchange from plant organs or small animals.

Standard meteorological stations usually measure only sunshine duration. The traditional instrument to observe this sunshine is the Campbell-Stokes sunshine meter. WMO abolished the world standard status of this sunshine recorder in 1989, as the process of evaluating the burns on its daily cards was both cumbersome and arbitrary. Instead, sunshine duration has been defined as the time during which direct radiation (on a plane perpendicular to the Sun's beam) is larger than 120 Wm^{-2}. This definition makes it possible to now use automatic sunshine recorders.

2.2   Temperature

Meteorological knowledge about energy exchanges and transports at the surface has useful applications in agriculture. With this knowledge it is possible to manage the climate near the ground by influencing the moisture status and temperature of upper soil layers. Examples: Ploughing the soil, mulching, shading and wind breaks. Some of the radiation striking a surface will be conducted into the ground. Heat energy is thus stored in the ground during the day, only to be released from the ground into surface at night. Therefore, ground acts as a "sink" of heat during the day and "source" of heat at night.

2.2.1   Air Temperature

Large quantities of energy are transferred between the surface and the air by the process of convection and the transfer (flow) determines air temperature. The mechanism of transfer is molecular diffusion rather than direct conduction. The daily oscillation of air temperature describes a sine curve, with the minimum normally occurring in the early morning hours around sunrise and the maximum occurring some time after peak solar and net radiation. The patterns of air temperature waves is not necessarily regular when examined on the basis of individual days, especially in climates that feature frequent frontal passages, irregular cloudiness, or strong advection of sensible heat generated in other regions. The afternoon temperature lag is a result of the balance between incoming and outgoing radiation. From sunrise on, a considerable amount of radiant energy is required to heat the soil and crop, which are cool at that time. Until these surfaces become warm relative to the air above, no net sensible heat flux to air occurs. The reason for annual lag is no different than for the daily lag. During spring and early summer, a large portion of the incident solar energy flows into the soil, which has reached its lowest temperature by the end of winter. As that portion of energy flux decreases and the soil becomes warm relative to the environment, more energy is converted to sensible heat.

2.2.1.1 *Measurement of Temperature of Air*

Small and simple radiation screens, some of which are aspirated when this does not destroy temperature profiles, are useful for special field work. High outside reflectivity, low heat conductivity, high inside absorption and good ventilation are desirable requirements in the construction materials and design. An idea of the radiation errors can for example be determined by simultaneous, replicated observations with the ventilated Assmann psychrometer at the hours of maximum and minimum temperature.

The most common thermometers for standard observations in air are those generally called *differential expansion thermometers*, which include liquid-in-glass, liquid-in-metal and bimetallic sensors. Because of their sizes and characteristics many of these instruments are of limited use for other than conventional observations. However, spirit-in-glass, mercury-in-glass and bimetallic sensors make useful *maximum and minimum* temperature measurements. When temperature observations are required in undisturbed and rather limited spaces, the most suitable sensors are electrical and electronic thermometers which permit remote readings to be made.

Resistance thermometers are metallic annealed elements, generally of nickel or platinum, whose electrical resistance increases with temperature; readings are made with appropriately scaled meters such as power bridges.

Thermocouples are convenient *temperature sensors* because they are inexpensive and easy to make. Those most frequently used in the environmental temperature range are copper-constantan thermocouples, which have a thermal electromotive force response of about 40 μV $^{\circ}C^{-1}$. This relatively weak response can be increased by connecting several thermocouples in series or using stable solid state d.c. amplifiers. Thermocouples are excellent for measuring temperature differences between the two junctions, *e.g.* dry and wet bulb temperatures, or gradients. When they are used to measure single temperatures or spatial average temperatures (such as surface temperatures, using thermocouples in parallel) there is always a need for one junction to be at a known steady reference temperature.

Thermistors are temperature sensors increasingly used in agricultural and animal micrometeorology. They are solid semi conductors with large temperature coefficients and are produced in various small shapes such as beads, rods and flakes. Their small size, high sensitivity and rapid response are valuable characteristics, offset however by their lack of linear response (less than metallic resistances) in the resistance-temperature relationship. Additional components are therefore required to achieve linear output.

Diodes and transistors with a constant current supply which provide outputs much higher than 1 mV $^{\circ}C^{-1}$ have been used to construct sensitive and accurate thermometers for application in plant environments.

Infrared thermometers are extensively used in micro meteorological studies.

In animal micrometeorology special and relatively simple instruments have been used to simulate the cooling power of the air or the heat load over the homeothermic animal body. *Kata thermometers* are spirit-in-glass thermometers with a rather big bulb of accurately determined area. With these, the time required for a fixed amount of cooling after the thermometer has been warmed above body temperature is measured. Such a reading is an index integrating the cooling effect of temperature and wind.

A *black-globe thermometer* is a blackened copper sphere commonly 15 cm in diameter, with a thermometer or thermocouple inserted. When a black globe thermometer is exposed in the open or under a ventilated shelter, the effects of different radiation fluxes are integrated with that of convective heat (wind and air temperature). Installed inside closed barns or stables, under still air conditions, it gives the average radiant temperature of soil, roof and walls at equilibrium.

Particular metadata for air temperature measurement are

(a) Height of sensor

(b) Description of employed screens (dimensions, material, ventilation).

2.2.2   Soil Temperature

The soil mantle of the earth is indispensable for the maintenance of plant life, affording mechanical support and supplying nutrients and water. The soil stores energy during the warm seasons and releases it to air during the cold portions of the year. The flux of heat into and out of the soil is a process of conduction.

Crop plants live in two media viz., air near the ground (stems and leaves) and in upper layers of the soil (roots). The soil temperature is critical for seed germination, root elongation and tuber development etc. It also effects decomposition of organic matter. Optimum soil temperature for different crops in degrees centigrade are wheat (15-18); sugar beat (16); field beans (21); maize (20); cotton(21); sorghum and soybean (25) and tobacco (29). Soil temperature is also important for economic parts of plants below the soil surface. Example: The tubers of potatoes develop best at 18; growth reduced at 20-27 and stopped at 2 degrees centigrade.

2.2.2.1   *Measurement of Soil Temperature and other Bodies*

Soil

Although the thermocouple must be of a sturdy construction, provided that presence of the sensor does not affect the temperature being measured. Mercury in glass type soil thermometers are frequently used. For measurements of the soil temperature at shallow depths these thermometers are bent in between angles 60° and 120° for convenience. At greater depths lagged thermometers are lowered into tubes. Care should be taken to see that water does not enter the tubes. Alternatively, shielded thermocouples or thermistors can be used. Temperature of deeper soil layers can be measured with glass thermometers, thermistors, thermocouples, diodes and platinum resistance thermometers when good contact is made with the soil.

In cold and temperate climates where the soil is often deeply frozen and snow covered, when continuous soil temperature records are not available or when many observing points are needed, different types of *snow cover and soil frost depth gauges* can be used. These instruments generally consist of a water filled transparent tube, encased in a plastic cylinder that is fixed in the soil. The tube is periodically removed from its plastic casing to determine the depth to which the entrapped water is frozen. If the fixed cylinder extends sufficiently far above the soil surface it can be used as a snow cover scale provided the exposed part is graduated.

For the soil surface temperature, non contact infrared thermometers are preferable as long as emissivity is known and again the presence of the sensor does not affect the temperature being measured by shading or otherwise influencing the natural radiation balance.

Particular metadata for soil temperature profile measurements are

(a) Instrument depths

(b) Regular specifications of the actual state of the surface.

Other Bodies

Like the soil, plant parts such as leaves, stems, roots and fruits have mass and heat capacity. The temperature of all these organs, can be measured with platinum resistance thermometers, thermistors, thermocouples, infrared thermometers and diodes, if the instruments do not influence the energy balance of those bodies. To measure their surface temperatures and those at the outside surface of animals, small contact sensors like thermocouples and thermistors or non-contact methods should be used.

In animal micrometeorology special and relatively simple instruments have been used to simulate the cooling power of the air or the heat load over the homeothermic animal body. Kata thermometers are spirit in glass thermometers with a rather big bulb of accurately determined area. With these the time required for a fixed amount of cooling after the thermometer has been warmed above body temperature is measured. Such a reading is an index integrating the cooling effect of temperature and wind.

A practical thermo-anemometer is the heated globe anemometer, which provides a reasonable value of the cooling power of air motions in climatic chambers and other indoor environments. It is constructed with a chrome plated sphere of 15 cm diameter heated by a nichrome wire that can receive a variable power input. Several thermocouples in parallel with one junction fixed internally to the globe wall measure the temperature of the globe wall. The voltage of the heater is regulated to give a differential air globe temperature of 15 °C. The power needed to maintain a steady temperature is a function of the ventilation. However, a correction factor for thermal radiation of walls, ground and roof may be required if these are significantly hotter than the air.

2.2.2.2   *Effect of Shading on Soil Temperature*

The practice of shading (interposing some type of screening against solar radiation) is employed in crops. Provision of vegetative cover also has macro scale implications as a protection against wind and water erosion. The influence of shading on the microclimate is rather complicated. A considerable fraction of total global radiation is reflected by the surface of trees providing shading or roots. For trees with a closed canopy, reflection factors vary from 16-37 percent. The reflection of dried leaves is higher than that of (the same leaves) green leaves and even when dried leaves are made wet. The non-reflected part of the global radiation is partly absorbed by trees providing shading and partly transmitted. If shading is provided by a roof made of dead material, most of the absorbed energy will be transformed into thermal energy. Therefore, the day time microclimate of shaded plants is characterized by, lower temperature of air and soil; lower light intensity; poorer light quality for photosynthesis and reduced evapo transpiration. During the night, however, the same shading will restrict the net outflow of terrestrial radiation, giving rise to, higher temperatures of air and soil and increased evapo transpiration. Soil temperature and moisture are very important variables that control plant growth and thus the ultimate yield of crops. Seedlings may not germinate unless the temperature is near an optimum range. However, soil moisture is also important, with excessive levels resulting in fungal attack and low levels resulting in continued dormancy. The soil surface may be chosen or modified to create a favourable environment for a crop. The slope or aspect of a soil may be chosen by picking an appropriate field (such as on the slope of a hill) or it may be modified in a localized manner by controlling the direction of ploughing. The slope and aspect of a field may be chosen such that the field has enough light throughout the day, has more radiation in the morning or afternoon than in the other half of the day. Ploughing may be oriented such that the furrows have a relatively constant temperature during the day, or one side has a greater temperature in the morning than the afternoon or vice versa.

2.2.2.3 *Effect of Surface Colour on Soil Temperature*

Differently coloured soil surfaces have different coefficients of reflection of shortwave solar radiation. So, different proportions of solar energy are absorbed by the surface. Net radiation of the soil surface can be changed by altering surface colour. Infrared albedos are always around one (1). So, controlling long wave exchange is not possible. Therefore, application of coloured surface dressings is used in agriculture to warm the uppermost layers of the soil or to prevent high temperatures which are dangerous for plants and seedlings. When the whole spectrum is concerned of marked differences in reflection of diffuse radiation from natural surfaces only occur between the extreme conditions. *Example*: Black cotton soil, or soil covered with root or carbon balance on the one hand; and on the other hand, light coloured, dry sand, chalk and in particular fresh snow. Basically, the effective colour of a soil changes with its moisture content and surface dressings make it possible to increase or decrease soil temperatures.

Table 2.1 Relationship among soil, albedo and variations in soil temperature

S. No.	Soil	Colour	Albedo	Temperature reductions	
				at (Surface)	at (5 cm deep)
1.	Grass covered soil	Green	0.32	26.5	--
2.	Fresh chalk	White	1.0	14.2	6.8
3.	Cotton soil (control)	Almost black	0.16	0	0

The temperature reduction measured over grass vegetation is greater than that from (white) chalk powder for two reasons. First, the vegetation is actually shading the soil surface and second the grass transpiration requires more energy than soil radiation. A thin layer of white powdered lime applied on black cotton soil decrease the surface temperature by about 15 °C and at a depth of 20 cm the soil temperature decrease by 5 °C. A dressing of magnesium carbonate reduce the daily maximum temperatures of the upper layers of the soil by 10 °C. In temperate climates (that have a cold season) dark materials are applied to the soil to increase the soil temperature in springtime for early germination. When vertical walls painted in dark and light colors; against the dark wall (which was warmer) the growth of tomato stem was stronger. However, against the light coloured wall the tomato crop resulted in more buds and increased yields (due to reflection of the short-wave radiation).

2.2.2.4 *Effect of Mulching on Soil Temperature*

Mulching is the application or creation of soil cover which reduces vertical transfer of heat and water vapour. Mulches also change surface albedo, thermal conductivity and thermal capacity. Most important mulches are loose layer of top soil (harrowing produces a mulch, apart from other effects); cut off gathered vegetable material such as grass, weeds, straw, tree leaves; redeployed surface material, such as stubble, litter, and stones; manufactured materials, such as paper, plastics and reinforced aluminum foil. The growing practice of employing artificial mulches by utilizing plastics requires special mention. The use of relatively rigid sheet material deliberately supported by stubble or pegs a few centimeters from the soil surface (with a thin layer of air between the plastic cover and the ground) act as glass houses. Prolonged snow cover also operates in many climatic situations as a mulch, protecting the sub soil from intense cold. It also has the practical advantage over other mulches as it applies itself in early winter and removes itself in springtime. The thermal conductivity of mulches is low. in case of heat conducting covers, the aluminum foil, the non convecting air layer below the cover acts as heat flow barrier. This results in daily and yearly surface temperature variations transmitted poorly downwards, so the subsoil is sheltered from short term temperature excesses, from midday or summer heat, and from night time or winter cold. Evaporation from the soil is reduced by mulching because of decreasing availability of vaporization energy below the mulch. Moreover, most mulches are an effective barrier to the upward flow of water vapour. If the mulch is a loose layered soil, resulting from cultivation, another reason for evaporation decrease is the rupture of capillary connections inhibiting the upward motion of deep soil water. If incoming radiation cannot be well transmitted into the soil, it heats up the surface. Similarly, if at night the outgoing surface radiation cannot be replenished from below, the surface cools more. As a result, the diarual temperature variation just above a mulch is greater than that would be on the surface of the bare soil. This difference also depends on the albedo of the mulch. One of the disadvantages of mulches is that, as soon as young plants extend their tips above the mulch, these suffer extremely large daily temperature variations. At that growth stage, removal of the mulch may be necessary. The choice of a mulch, its conductivity and its albedo, certainly requires much thought. Black plastic and hay heat up highly, the plastic because of low albedo, the hay because of low conductivity. Hay and paper have a significant soil cooling effect. All mulches reduce evaporation, even hay. A new aluminum film cover has a high albedo and behaves like paper mulch; old aluminum has a very low emission (0.5). White polythene sheet leaves soil temperatures nearly unchanged, while black and transparent polythene results in soil temperature increase.

There are few disadvantages of mulches. It can be quite harmful when the activity of soil microbes is restricted, either by lowering of temperature or by the depletion of nitrogen. In the latter case, bacterial activity is concentrated more in the organic mulch than in the soil itself. Mulches can alter the composition of microfauna. A major negative effect of mulching is the increase of frost risk to crops. In practice, the first priority in reducing frost danger in a plantation or orchard is to remove loosely packed vegetation litter where such a layer conducts heat very high. Other frost preventive action where consists of increasing the contact between soil particles by compaction, in order to increase the conductivity for soil heat flux. Soil compaction should perhaps be seen as an inversion of mulching. In a European climate, the loosening of the upper soil layer to a depth of 2 cm will lower the night minimum surface temperature by 2 °C, for moist soil and 3 °C for dry soil.

2.2.2.5   Effect of Micrometeorological Technics on Soil Temperature

The growth of roots and other plant parts may be related to soil temperature. When the soil temperature is less than optimum value, increasing the soil temperature may result in increased yield. For economic reasons, it is suggested that the waste heat from power plants and other industrial plants be used for agricultural purposes. Artificial heating by direct electrical methods can also be used. Yield increases of up to 100 % have been predicted, especially when the soil heating is early in the growing season. In Keith Mayberry, Imperial County, the growers mainly use the mid trench method for winter melon production. The mid trench methods are also used for cucumbers and peppers in San Diego Country. Growers make a trench in the bed about 5 to 6 inches deep and plant in the trench. The trench is covered with either clear plastic or embossed (semi-opaque) plastic. Studies showed that east-west tunnels work best. These are winter crops and a south facing slope in the tunnel is much warmer. The embossed plastic is cooler during daylight and warmer at night. They had a frost problem with clear plastic. The embossed plastic maintained temperatures 2 °F to 3 °F warmer at night. Problems occur with weeds, knowing when to ventilate the plastic and knowing when to remove the plastic. Removing the plastic without ventilating first causes severe wilting because of the large humidity change. Ventilation is accomplished by cutting holes in the plastic every 2 to 3 feet to allow humidity to escape more slowly. Cutting the plastic often causes post problems because insects can enter the tunnels. No ventilation restricts root growth. Weeds are a serious problem with most plastic mulches. There is no way to cultivate them. New herbicides are being investigated. New biodegradable plastics are available, but, the cost is about 50 percent higher and they have not been widely accepted.

A bridge type approach is used for tomatoes. In this approach two ridges are formed on top of the bed and the clear plastic forms a bridge over the seeded tomatoes. The plastic warms up the beds, improves soil moisture status and protects plants from birds. In Coachella Valley, black plastic is used over drip tubes for pepper production. The black plastic improves temperature and soil moisture and it reduces weed problems. The other big problem is how to dispose of the plastic. The new biodegradable plastic could help, but, they are too expensive. The biodegradable plastics also are good in that they self ventilate and remove themselves. Currently, non-biodegradable plastics are rolled up and burned.

In Harry Otto, Orange Country, the plastic mulches are used to grow strawberries as winter crop. The growers use clear plastic to warm soil mainly in October and November to encourage growth. However, for summer crop growers use white plastic to cool the soil through reflection. Also, plastic mulches minimize mould of strawberries by keeping off the ground. In case of bell peppers the farmers start with clear plastic in the spring to warm soil and encourage growth. Later change to black plastic for weed protection.

2.3   Atmospheric Humidity

The amount of water vapour in the air, which controls the flux of water to and from a surface, is measured by humidity. Humidity may be quantified in a variety of ways, including mass of water vapour per unit volume of air (absolute humidity), mass of water vapour per unit mass of air (specific humidity), mass of water vapour per unit mass of dry air (mixing ratio), vapour pressure (the partial pressure exerted by the water vapour molecules), the dew point temperature (the temperature to which air must be cooled under specific conditions to produce saturation) and relative humidity (vapour pressure in the air divided by the saturation vapour pressure evaluated at air temperature).

The most important measures of humidity involve the concept of saturation. When a flat surface of water is in contact with the atmosphere, some molecules will theoretically be evaporating from the surface while others will be condensing. This is because at any temperature, molecules are moving at a range of speeds. The slower gas molecules will be moving slowly enough to be attracted to the water surface and be captured, while the faster moving liquid molecules will be moving fast enough to escape the water surface and become a gas. At a certain temperature, an equilibrium will be reached where the number of molecules leaving the surface exactly equals the number entering the surface. A certain concentration (absolute humidity) of water molecules will be associated with this equilibrium. According to Dalton's Law of partial pressures, the concentration will create a particular saturation partial pressure or vapour pressure, known as the saturation vapour pressure in this case. Because the saturation vapour pressure is a function of temperature, one should always conceive of it as a function. When the concentration of water molecules surpasses the saturation concentration, more molecules will enter the surface than will leave it. A net flux of water vapour molecules to the surface will occur, creating condensation on the water surface. Thus, if the water vapour pressure in the air is slightly greater than the saturation vapour pressure, condensation on a flat pure water surface is expected. Super saturation occurs when the vapour pressure and concentration in the air is greater than the saturation vapour pressure and concentration. This may occur when there are insufficient condensation nuclei on which water condenses.

Relative humidity, the most common measure of humidity, is defined as the quotient of atmospheric vapour pressure divided by the saturation vapour pressure. By definition, saturation occurs when the relative humidity is 100% or 1.00. Super saturation thus occurs when the relative humidity is greater than 1.00 or 100% which is hypothetical. Relative humidity may be expressed in either manner, although it is usually expressed as a percentage. In most cases, near the ground, when the relative humidity is close to 100%, condensation will occur. The vapour pressure deficit is defined as the difference between the atmospheric vapour pressure and the saturation vapour pressure at air temperature.

Dew formation, or the condensation of water on a surface, from the air above, is only one of the ways in which plants appear 'wet' in the early morning. Another method is water of guttation which is defined as "The excretions of exudates from plants". A third process is that of distillation, where water evaporates from the soil and condenses on the plant canopy.

Lower humidity cause severe atmospheric drought. More heat units are required during the total growing period of the same variety sown in dry season than during the rainy season, even when the temperatures are the same between seasons. However, for sorghum, under high relative humidity (>80%) the duration of the total life cycle is shortened, while low relative humidity (<50%) tends to lengthen the total life cycle.

High atmospheric humidity has two beneficial effects on plant growth. First, many plants can directly absorb moisture from an unsaturated air of high humidity. Second, humidity may effect the photosynthesis of plant leaves. The effect of vapour pressure deficit on apparent photosynthesis of cotton leaves at 40 °C and at various high intensities and times of the day revealed that the photosynthetic rate increases with humidity, only at low light intensities but quite substantially at high intensities.

Most plants grow well under high atmospheric humidities, except when saturated air persists for weeks and completely stops transpiration. By varying relative humidity from 50 to 90 percent it was found that the number of flowerings of groundnut increased with humidity. High humidity at night is especially beneficial. Maize and tomato grown under high humidity not only gained more weight but also developed better root systems. Evapotranspiration rate decreases with the increase of humidity. Significantly higher rate of water use was reported at 40 percent relative humidity than at 95 percent humidity in the early stages of maize crop growth. During the dry summers, crops often exhibit a distinctly different growth response to showers that are accompanied by a high humidity than they do to irrigation that is accompanied by a low humidity. It was found that the size of raspberries increased after a very light shower, an increase that was not observed after irrigation. It was also established that other things being equal, the efficiency of water use increase with the humidity of air.

2.3.1   Measurement of Humidity

The most commonly used hair hygrometers and hair hygrographs may give acceptable values only if great care is taken in their use and maintenance. Besides standard *psychrometers equipped with mercury-in-glass thermometers, portable aspirated and shielded psychrometers and mechanical hygrometers*, many instruments have been developed to measure different aspects of air humidity. Since, the above mentioned routine instruments are bulky and inadequate for remote reading, they are unsuitable for many agrometeorological observations. For observations in undisturbed and small spaces, electrical or electronic instruments are used. The best method for measuring humidity distribution in the layers near the ground is also by using thermo electric equipment, unventilated thermocouple psychrometers being most suitable in vegetation. Ventilated psychrometers may be used for levels at least 50 cm above bare soil or dense vegetation.

For measuring relative humidity directly, *lithium chloride or sulphonated polystyrene layers are also used,* since the electrical resistance of these electrolytes changes with relative humidity. However, these electrolytic sensors become affected by air contamination and high relative humidity conditions and are therefore to be used with great care and frequent recalibration. For example *resistive polymer film* humidity sensors are increasingly used. Instruments are usually contaminant resistant and common solvents, dirt, oil and other pollutants do not affect the stability or accuracy of the sensor.

Electrical dewpoint hygrometers indicate dew point rather than relative humidity. For example, the *lithium chloride dew point hydrometer* measures the equilibrium temperature of a heated soft fiberglass wick impregnated with a saturated solution of lithium chloride. The temperature is linearly related to atmospheric dew point. However, the response of the instrument under low relative humidity conditions is not so good.

More expensive and complicated, but more accurate, instruments require that the air be sampled and delivered, without changing its water-vapour content, to a measuring unit. One such instrument, an illuminated condensation mirror, is alternately cooled and heated by a circuit energized by a photocell relay, which maintains the mirror at dew point temperature. *Infrared gas analyzer hygrometers* (IRGAs) are based on the fact that water vapour absorbs energy at certain wavelengths and not others. Two sampling tubes are also used to measure absolute values of water vapour concentration at two levels.

Single or double junction peltier psychrometers are extensively used for accurate measurement of water potential values in plant tissues and soil samples. They are generally based on the Peltier effect in chromel constantan junctions and the water potentials are derived from measurements of equilibrium relative humidity in representative air.

Particular metadata for any type of hygrometry are regular notes in the station logbook of maintenance activities, such as psychrometer wick replacement or cleaning of sensor surfaces. Moreover, it should be recorded if sensors are ventilated. Because so many different humidity parameters are in use, the metadata should specify not only the actually used parameters and units, but also contain information on the way in which the archived humidity data were calculated from original observations (conversion tables, graphs, small conversion programmes).

2.3.2   Soil and Grain Moisture

Time and space variation of soil moisture storage is the most important element of water balance for agrometeorology. Several instruments have been constructed to measure soil moisture variations at a single point but they avoid the variability of soils in space and depth. Several scientists described indirect methods of obtaining soil water content as "measurement of a property of some object placed in the soil, usually a porous absorber, which comes to water equilibrium with the soil". Blotting paper is popular here and they may be also useful for soil potential determinations, characterizing the water supplying power of the soil in representation of root hairs.

Subjective methods of estimating soil moisture have been used with satisfactory results in some regions where regular observations in a dense network are necessary and suitable instruments lacking. Skilled observers, trained to appreciate the plasticity of soil samples with any simple equipment, form the only requirement for this method. Periodic observations and simultaneous determinations of soil texture at depths, by competent technicians, allow approximate charts to be constructed.

The direct methods of soil water measurement facilitate implementation of easy follow up methods at operational levels. Gravimetric observations of soil water content have been in use for a long time in many countries. An auger to obtain a soil sample, a scale for weighing it, and an oven for drying it at 100-105 °C are used for the purpose. Comparison of weights before and after drying permits evaluation of moisture content which is expressed as a percentage of dry soil or where possible by volume (in mm) per meter depth of soil sample. Because of large sampling errors and high soil variability the use of three or more replicates for each observational depth is recommended. The volumetric method is useful to measure the absolute amount of water in a given soil and it has known volumes of soil sampled.

Tensiometers measure soil moisture tension, which is a useful agriculture quantity, especially for light and irrigated soils. The instrument consists of a porous cup (usually ceramic or sintered glass) filled with water, buried in the soil and attached to a pressure gauge (e.g., a mercury manometer). The water in the cup is absorbed by the soil through its pores until the pressure deficiency in the instrument is equal to the suction pressure exerted by the surrounding soil. Next to this direct measurement an indirect measurement of soil moisture tension can be obtained from electrical resistance blocks.

Electrical resistance blocks of porous materials (e.g. gypsum), the electrical resistance of which changes when moistened without alteration of the chemical composition, can be calibrated as a simple measure of soil moisture content. It was operationally used successfully.

Among *radioactive methods, the neutron probe* measures the degree to which high energy neutrons are thermalized in the soil by the hydrogen atoms in the water. It determines volumetric water content indirectly *in-situ* at specific soil depths using a pre designed network of access tubes. The neutron scattering and slowing method was until recently most widely used, relatively safe and simple to operate. The total neutron count per unit time is proportional to the moisture content of a sphere of soil of which the diameter is larger when the soil is drier. Soil moisture is measured with the *gamma radiation probe* by evaluating differential attenuation of gamma rays as they pass through dry and natural soils. This method generally requires two probes introduced simultaneously into the soil a fixed distance apart, one carrying the gamma source and the other the receiver unit.

Time Domain Reflectometry (TDR) determines the soil water content through measuring the dielectric constant of the soil, which is a function of the volumetric water content. It is obtained by measuring the propagation speed of alternating current pulses of very high frequency (>300 MHz). The pulses are reflected at in homogeneities, either in the soil or at the probe soil interface and the travel time between the reflections is measured. From the travel time the dielectric constant is determined and in that way the volumetric water content of the soil. As with neutron scattering, this method can be used over a large range of water contents in the soil. It can be used directly within the soil or in access tubes. Compared to the neutron scattering method, the spatial resolution is better, calibration requirements are less severe and the cost is lower.

In agriculture another important measurement needed is the moisture content of grains which influences viability and general appearance of the seed before and after storage. It is important to know the moisture content immediately after harvest, prior to storage, shipment and after long periods of storage. The methods for measuring moisture content are generally classified as reference methods, routine methods and practical methods. The phosphorous pentoxide (moisture is absorbed by this chemical) and Karl Fisher method (water is extracted from seed using a reagent) come under reference methods. The "Oven dry method" is categorized as routine method in which the seed moisture is determined by removing the moisture from the seeds in an oven. Among the practical methods the moisture content determination by using samples in infrared moisture meters is easy as compared to others.

2.3.3   Leaf Wetness and Dew

The very large number of instruments that have been developed for the measurement of dew or duration of leaf wetness indicates that not even a moderately reliable method has yet been found. Two main categories of Leaf Wetness Duration (LWD) sensors being used are:

(i) Mechanical sensors with recorders

(ii) Electric sensors, which exploit the conductivity variation as a function of wetness.

In addition to electric conductivity measurements of dew (variations on both natural and artificial surfaces), principles of mechanical dew measurement are:

(i)   Modification of the length of the sensor as a function of wetness

(ii) Deformation of the sensor

(iii) Water weighing type (dew balance recorder)

(iv) Adsorption on blotting paper, with or without chemical signaling.

There is also visual judgment of drop size on prepared wooden surfaces (Duvdevani).

Porcelain plates (Leick), pieces of cloth and other artificial objects can share in any dew fall or distillation occurring on a given natural surface. Unless they are more or less flush with that surface and have similar physical properties (surface structure, heat capacity, shape, dimension, flexibility, color and interception) they will not indicate reliably the amount of dew that the surface receives. If exposed above the general level of their surroundings, as is normal with Duvdevani blocks and usually appears to be the case with more refined "Drosometer" devices, their behavior will diverge from that of the surface below and the observed amounts of dew may bear little relation to the dew on adjacent natural surfaces.

Weighing type instruments, modified hygrographs with a hemp thread instead of a hair bundle and systems with surface electrodes that connect when the surface is wet, all have their problems. The surface electrode instruments are the simplest to read, but again do not measure real leaf wetness, because the sensor is again a fake leaf, with different heat capacity.

2.4 Wind

Wind effects plant growth in at least three significant ways viz, transpiration, carbon dioxide intake and mechanical breakage of leaves and branches. Transpiration increases with wind speed up to a certain point, beyond which either it does not increase or it decreases slightly at high wind speed. Wind exerts a much greater influence on cuticular transpiration than on stomatal transpiration. Therefore, plants with higher cuticular transpiration (hydrophytes) show an appreciable increase in the transpiration resulting from the action of the wind. Under natural conditions, the effect of wind on transpiration vary according to roughness as determined by the surface configuration. In general, the effect is greatest for an isolated tall plant. An increase in the transpiration rate of coconut palm in full sunlight of about 100 percent with a wind speed at five miles per hour was estimated as much more than coconut palms grown in an acre of land. For any other crop also, if the cover is complete and the surface is more or less even and smooth, then the effect of wind on transpiration is usually small.

In addition, wind is important to agriculture in both direct and indirect ways. Extremes in wind speed can result in the lodge of crops. Changes in wind speed indirectly affects crop surfaces by changing the resistance and thus controlling the fluxes of heat, water vapor, carbondioxide and pollutants. Wind is also important to the application of pesticides, fertilizers and irrigation water. Wind in the atmosphere near the ground is very unsteady. Gusty conditions are frequent representing large scale turbulence. The more unstable the atmosphere the more turbulent it is with enhanced mixing. It is the turbulence that is the primary factor determining aerodynamic resistance.

2.4.1 Measurement of Wind

Wind speed and direction measured with standard instruments under standard exposure are fundamental requirements of the science of agricultural meteorology. The most common routine observation is the wind run, providing an average over the measuring period. That period should be at least ten minutes for smoothing out typical gustiness and at the most an hour because surface wind has a very pronounced diurnal course. However, different instruments are used when it is necessary to observe the more detailed structure of air motion, e.g., in agricultural meso and micrometeorological studies. In such cases wind speeds are measured with cup anemometers of high sensitivity at low velocities or with electrical thermo anemometers or sonic anemometers.

Sensitive *cup anemometers* are the most common in routine and research use, measuring all wind components with an angle of attack with the horizontal smaller than about 45°. The best have a low stalling speed (threshold of wind speed below which the anemometers does not rotate) of about 0.1 m s^{-1}, because friction loads have been minimized. The rotation produces an electrical or phototransistor signal which is registered by a recorder or counter. Such transducers also allow separate recording of gustiness.

Sensitive *propellers,* if mounted on a vane, could be an alternative for cups but are mainly research instruments. *Pressure tube anemometers* on a vane are reliable but, so unwieldy that they are disappearing in favor of smaller instruments. A new instrument for horizontal wind speed and direction measurement is the *hot disk anemometer*, which has the advantage that it has no moving parts. For steady wind direction measurement, wind vanes must have fins whose height exceeds their length.

Sonic anemometers, sensing the transport speed of sound pulses in opposite directions along a line, so being totally linear, respond quickly enough to measure turbulence and have become useful in flux measurements in research. However, they cannot be used in small spaces and their calibration shifts in wet weather.

For research of wind speeds in restricted spaces, such as crop canopies and surfaces, several kinds of thermo anemometers are used. The *hot wire anemometer* is an electrically heated wire, of which the heat loss is a function of the airspeed at normal incidence to the wire. It is particularly useful for low speed winds but very fragile and in polluted surroundings it loses its calibration, therefore, it cannot be used operationally. Because of the dependence of wire heat transfer on wind direction, crossed wire sensors can be used to separate the wind components in turbulent motion.

Hot bead anemometers have heated beads, of which heat transfer is less dependent on wind direction but has a slower response. Thermocouples or thermistors sense differences in temperature between heated and non heated beads, which differences are a function of the wind speed. Shaded Piche evaporimeters have also been used as cheap interpolating and extrapolating ancillary anemometers in agroforestry under conditions of not too high turbulence and low temperature and humidity gradients.

Particular metadata for wind measurement are

(a) Response time of instruments

(b) Sensor height

(c) Exposure, i.e., adequate description of surrounding terrain and obstacles

(d) Type of anemometer signal, its transmission and its recording

(e) Sampling and averaging procedure

(f) Unit specification (m/s, knots, km/hr, or some type of miles per hour).

2.5   Atmospheric Pressure

The lower pressures experienced as altitude increases have important consequences for high altitude plant life. At high altitudes and low atmospheric pressures the solubility of carbon dioxide and oxygen in water are reduced. Some plants show stunted growth at higher altitudes as concentrations of oxygen and carbon dioxide reaches low. Plants with strong root system and tough stems can live under increased wind speeds at low pressures in high altitude areas. It is usually adequate to know the altitude at which an event takes place, but, if air density is needed within $\pm 5\%$ then pressure variations have to be taken into account. Usually, a station will record pressure as part of their data for climatological work.

2.5.1   Measurement of Atmospheric Pressure

Analyzed pressure fields are useful in agricultural meteorology. These pressure fields must be accurately defined because all the subsequent predictions of the state of the atmosphere depend to a greater extent on these fields. In mercury barometers the pressure of the atmosphere is balanced against the weight of the column of the mercury, whose length is measured using a scale graduated in units of pressure. Of the several types of mercury barometers, fixed cistern and Fortin are most common. For the purpose of comparison, pressure readings may need to be corrected for ambient air temperature.

In electronic barometers, transducers transform the sensor response in to a pressure related electrical quantity in the form of either analogue or digital signals. Aneroid displacement transducers, digital piezo resistive barometers, cylindrical resonator barometers fall into this category. Calibration drift is one of the key sources of error with electronic barometers. Therefore, the ongoing cost of calibration must be taken in to consideration when planning to replace mercury barometers with electronic ones.

Aneroid barometers have the advantage over conventional mercury barometers that they are compact and portable. Another important pressure measuring device is the Bourdon tube barometer. It consists of a sensor element (aneroid capsule), which changes its shape under the influence of pressure and transducers which transform the change into a form directly usable by the observer, such as on a barograph. The display may be remote from the sensor.

2.6 Rainfall/Precipitation

The aspects of the hydrologic cycle, involving the soil, are important in agricultural meteorology, for the reservoir of water in the soil permits transportation to continue and plants to survive between rains. The precipitation is divided into components according to how the water returns to the atmosphere. A certain fraction of the precipitation is intercepted by the plant canopy and evaporated without reaching the soil surface. Of that precipitation which reaches the soil surface, part infiltrates into the soil and part runs off over the surface of the soil, eventually finding its way to a stream. The water which enters the soil may return to the surface, either through the soil or through the transpiration stream of a plant and be lost to the atmosphere by evapo transpiration. The water which is not evaporated may be divided between that which percolates downward through the soil in response to gravity and joins the ground water and that which drains down slope to reappear as surface water at a lower elevation. This later flow is designated as groundwater runoff or base flow. It is this flow which feeds springs and maintains the flow of streams between storms.

Precipitation

Precipitation may occur in the form of rain, snow, hail, sleet or dew. The distribution of precipitation is seldom more than approximately uniform in space and is never uniform in time.

2.6.1   Measurement of Rainfall/Precipitation

WMO Technical Notes provide information and guidance concerning instruments such as rain gauges and totalisators, rain recorders (float and tipping bucket types) and snow gauges. For some purposes, great precision in rainfall is needed, for example, in classifying days as either "Wet" or "Dry" for insurance claims or when only rough ideas are wanted of accumulation of rainfall over agricultural fields over an ongoing season compared to the same period in earlier years, a topic most farmers will be interested in. The same applies in school (agricultural) environmental science teaching. In Mali, the National Meteorological Directorate is of the opinion that farmers need to have a means of measuring rainfall if they wish to derive full profit from the agrometeorological information disseminated by rural radio, and farmer rain gauges are locally manufactured.

Hailfall observations cannot be automatized, because the only useful observation method so far is employment of a network of hail pads. When cost is important, including needs for high measuring densities, rain gauges smaller in size than the normal standard are employed, but, these are unsuitable for snow. Sometimes these are made of plastic and shaped like a wedge, sometimes they are plastic or other "Can" like receptacles. Commercially, the former are often called "Rain gauges according to Diem" or "Farmer rain gauges", the latter, if from plastic, "Clear view raingauges". Inexpensive rain gauges and small size totalizer rain gauges are used for studying the small scale distribution of precipitation, as in limited meso climates, forest or crop interception shelterbelt effects.

In addition to routine rainfall measurements agricultural practices need the amount, duration and intensity of precipitation at the time of floods and related disasters. As the severe weather systems affecting coastal areas originate in seas and oceans, the ocean based data collections through ships and buoys are necessary. Also, installation of automatic weather stations that meet the necessary criteria will help monitoring and early warning coastal zones about hazardous weather. However, capability for disseminating the weather data, topography and vulnerability of the coastal zone to severe weather decide on the weather station network requirements. In vulnerable coastal zones a dense network of observations is required to diagnose and plan in advance the mitigation of weather related hazards.

Radar, sometimes together with satellite remote sensing, is increasingly used to estimate both point and area of rainfall from characteristics of cloud structure and water content. These data complement the surface raingauge networks in monitoring and mapping rainfall distribution, but, it is essential that representative actual observations at the surface are used when taking decisions on the track of the storm for forecasting purposes. Such derived rainfall data need ongoing intensity calibration.

Particular metadata for precipitation measurement are:
  (a) Raingauge rim diameter and rim height above ground

(b) Presence of a nipher screen or some other airflow modification feature

(c) Presence of overflow storage

(d) Manner, if any, of dealing with solid precipitation (heating, snow cross).

2.7   Evaporation and Water-Balance

Standard instruments used for measuring the different elements of the water balance for climatological and hydrological purposes (screened and open pan evaporimeters, lysimeters) are also employed in agricultural meteorology.

2.7.1   Evaporation

While it is possible to estimate actual or potential evapo transpiration from observed values of screen or open pan evaporimeters or from integrated sets of meteorological observations, more accurate, direct observations are often preferred. Actual evapo transpiration is measured by using soil evaporimeters or lysimeters, which are field tanks of varying types and dimensions, containing natural soil and a vegetation cover (grass, crops or small shrubs). Potential evapo transpiration (PET) can be measured by lysimeters containing soil at field capacity and a growing plant cover. A surface at almost permanent field capacity is obtained by regular irrigation or by maintaining a stable water table close to the soil surface. A strict control must be kept of infiltration from rainfall of excess water supply. The reliability of observation by lysimeters depends upon the conditions at the instrumental surface and below it being very similar to the conditions of the surrounding soil.

Among the different lysimeters the most important ones for agricultural applications are Thornthwaite lysimeters (drainage type), Popoff lysimeters (combined drainage and weighing type) and weighing lysimeters and hydraulic lysimeters (more robust weighing type). Lysimeters are used for measurement of evaporation, transpiration, evapo transpiration (ET), effective rainfall, drainage and chemical contents of drainage water, to study climatic effects of ET on the performance of crops. Lysimetry is one of the most practical and accurate methods for short term ET measurements, but, a number of factors cause a lysimeter to deviate from reality viz., changes in the hydrological boundaries, disturbance of soil during construction and conduction of heat by lateral walls.

Atmometers or "Small surface" evaporimeters are also still in use, among which the cheap Piche can be utilized anywhere in meteorology and agriculture if the physics are well understood. Shaded Piche evaporimeters were used to replace humidity and wind speed data in the aerodynamic term of the Penman equation.

Devices for measuring net radiation, soil heat flux and sensible and advected heat are needed in energy budget methods, while continuous measurements of wind speed, temperature and water vapour profiles are needed for the aerodynamic method. When adequate instrumentation facilities and personnel are available, it is possible to compute actual evapotranspiration using energy balance or mass transfer methods. Certain semi empirical methods which require relatively simple climatological measurements to provide estimates of PET are often of little value when evaporation is limited by water supply.

Microlysimeters are very small lysimeters, for soil evaporation measurements, which can be put into the ground and used for short times, such that disturbance of the soil boundary condition does not appreciably effect evaporation from the soil. Precautions and measuring protocols must be followed.

Particular metadata for pan evaporation are:

  (i)   The pan dimensions and rim height
  (ii) Any employment of pan defense against thirsty animals (e.g., wire netting).

2.7.2 Irrigation

Water balance studies are incomplete without proper reference to different methods of irrigation because water with acceptable quality is becoming a more and more scarce resource for agriculture. Measurements and calculations particularly include soil moisture conditions, water use efficiencies and water flow conditions in canals of different dimensions, including the smallest field channels.

2.8 Weather- Soybean (Glycerine max (L.) Merr)

The following is the original work of Dr. Murthy, the author of the book for his Ph.D.

2.8.1 Introduction

Food for all and freedom from the scourge of malnutrition is the goal for development planning in India. Oilseeds and pulses provide fats and proteins which are essentially linked to the staple cereal food ingredients of the country. These crops account for 14 and 12 percent of cropped area and production of 22 and 15 million tones respectively. The country would need around 30 million tones of pulses by 2020 A.D. The requirement of oilseeds would be more than the present estimates as the population of India is expected to go upto 1280 million by 2020 A.D. At the present rate of production it is extremely difficult to meet the dietary standards prescribed by Indian Council of Medical Research for oils, fats and proteins. So, India has to make rapid and unprecedented strides to sustain the needs of burgeoning population. Soybean (*Glycine* max (L.) Merr.) has the capacity not only to meet this challenge but also to revolutionize rural economy and uplift socio-economic status of farmers. Soybean is one of the 9 oilseeds crops unique in its encapsulate 20% oil and 40% proteins in addition to vitamins, minerals and essential aminoacids. This is also cost effective alternative to ameliorate pulse protein deficiency in the country. The need for its niche in the agricultural scenario in India is not a new vision. It is the dream of Mahatma Gandhi - The Father of the Nation - as long ago as 1930. He stated *"I advise popularization of soybean in India for its qualities to alleviate protein deficiency among masses at low cost"*. It is the fundamental responsibility of every Indian scientist to cherish this dream and work for food security.

With coverage of 5 million hectares and production of 4.8 million tonnes soybean has recently occupied important place in agricultural and oil economy of not only India but also the world. About 90 percent of suitable cultivated area and production in the country is mainly confined to Madhya Pradesh, Rajasthan and Maharashtra. This indicates a wide gap to be urgently abbridged by its diffusion in all other parts of the country. Sporadic research attempts provided optimistic opportunities of the intensification. Extensive research is warranted to study the adaptability and stability of this crop to varying agro ecological situations for authenticated recommendations for intensification or diversification with less economical crop. The day to day weather conditions are the most potent natural inputs that determine the adaptability of a crop and its productivity. It is time to change the minds of the peasants and give this crop its due share in their fields to bring prosperity for their hard working families.

Indian agriculture provides a unique perspective of economic development which has overcome the problem of food scarcity. Due to this, agriculture is recognized as the basis of our economy. But, the year to year variation in agricultural production is huge compared to other sectors. Added to this, the cost of inputs like fertilizers, pesticides, irrigation, and labour are increasing alarmingly. So far agricultural scientists have exploited all the resources including soil to the fullest extent for maximum crop yields. There are no chances to further exploit these inputs to meet the future demands of food. The only alternative left is using "Weather" as an input, which has a significant impact on crop growth and development. Being a non monetary "Input" in crop production, weather is simultaneously one of the "Factors" that effect the way a crop will respond to other inputs. Weather is also one of the key "Components" that controls agricultural production on a wider scale. Further, the crop is known to modify the weather within its canopy and the soil underneath it. This is further influenced by the cultural and management practices. Therefore, an array of genotypes are to be vigorously subjected to research on their relationship with varying climatic variables between and within different seasons to cull out candid recommendations for its successful husbandry to cherish the 85 year long dream of the *Father of the Nation* which is left behind and yet unaddressed completely. The objectives of the investigation are as follows:

- To determine the optimum date of sowing and to evaluate the most suitable soybean genotype
- To evaluate the crop productivity and resource use of soybean under varying environmental conditions
- To examine the effect of macro and micro meteorological factors on soybean crop production
- To develop growth and yield models based on minimum weather parameters and physical principles.

2.8.2   Material and Methods

The experiment was conducted at the Students Farm of the Agriculture College of Acharya N. G. Ranga Agricultural University head quarters at Rajendranagar, Hyderabad. The soil was alfisol in texture. Its reaction was neutral with 7.25 pH. The EC was normal (0.13 dsm^{-1}). The available nutrient status indicated that the soil had a poor status with 159 kg N ha^{-1}

but medium in available phosphorus with 40 kg P_2O_5 ha^{-1} and Potassium with 210 kg ha^{-1}. Soybean was raised during rabi seasons in 1996 and 1997, summer 1997, and Kharif 1997. The layout of the trial was a randomized block design. The treatments were a combination of five soybean genotypes and four sowing dates at 20 days interval in each replication. These were replicated thrice. The analysis of variance was subjected to the technique of 5 × 4 factorial approach. The selection of genotypes was biased to include three semi determinates two determinate types with extreme variation in maturity duration ranging from 70 to 120 days.

All cultural practices were normal. Fertilizers were applied 40, 60 and 30 kg ha NPK through urea, single super phosphate and muriate of potash respectively. The nitrogen fertilizer was split applied twice, half of it was applied along with the entire quantity of phosphorus and potash at sowing and rest was applied 30 days later. A brief descript of the meteorological observations recorded in the investigations are as follows:

Air temperature - maximum and minimum air temperature were recorded at 4′ height in the Stevenson screen.

Rainfall - This weather element was recorded with the rain gauge installed near the experimental site.

In addition to the above weather elements which were recorded near the experimental site, the open pan evaporation, wind speed and direction and sunshine hours were monitored from the adjacent observatory (< 1 km) from the experimental site.

2.8.3   Micrometeorological Studies

Measurement of the components of Photosynthetically Active Radiation (PAR)

The various components of PAR viz transmitted radiation, reflected radiation and intercepted radiation were measured at an interval of 7 days at solar noon, with the help of the line quantum sensor between 1130 and 1300 hours. To eliminate the effects of solar elevation, the measurements were made simultaneously at mid day. The line quantum sensor was connected to data logger and the value was recorded instantaneously from the data logger. Two values were recorded from each spot for accuracy and their average was considered.

Measurement of PAR (PARo)

For measurement of incoming PAR, the line quantum sensor was positioned facing up 30 cm above the top of the canopy and value was recorded for incoming PAR.

Measurement of Transmitted (TPAR)

Line quantum sensor was placed above the ground across the rows and value was recorded for TPAR.

Determination of Intercepted PAR (IPAR)

IPAR was calculated as follows

$$IPAR = PARo - TPAR$$

The IPAR values were recorded at an interval of 7 days from 21 DAS to harvest and they were converted into percentage IPAR.

Measurement of Reflected PAR (RPAR) and RPARc (canopy + soil)

Total reflected PAR was measured by inverting the line quantum sensor and holding 30 cm above the canopy across the row and spirit level adjusted and instantaneous value was recorded for RPAR. The percentage of RPAR was calculated as RPARc

$$RPARc \% = (R \, PARc/PARo) \times 100$$

Measurement of RPAR by soil (RPARs)

RPARs was measured by inverting the line quantum sensor and holding it 15 cm above the soil across the rows. Instantaneous value was recorded from the data logger. Percentage of RPAR by soil was calculated as under

$$RPARs \% = (RPARSs \, / \, PARo) \times 100$$

Determination of RPAR by Canopy

RPAR by canopy was determined by deducting the RPAR by soil from RPAR by canopy + soil

$$RPAR = RPARc - RPARs$$

Determination of Absorbed PAR (APAR)

APAR was worked out as follows

$$APAR = (PARo-RPARs) - (TPAR + RPARc)$$

Estimation of Radiation use Efficiency (RUE)

Radiation use efficiency was determined as

$$\text{RUE} = \frac{\text{Amount of drymatter produced}\left(\text{gm}^{-2}\right)}{\text{Amount of cumulative radiation absorbed}\left(\text{MJm}^{-2}\right)}$$

Cumulative APAR was calculated with the assumption that the daily total PAR is 0.50 of daily total solar radiation. Daily percentage of PAR was worked out graphically from the value of percent APAR determined at different growth stages. Further, it was multiplied by daily incident PAR to calculate APAR (MJm^{-2}).

The humidity was measured in the crop canopy and 50 cms above the crop canopy by using Assman Psychromteter in all the treatments at 7 days interval of crop growth. Micrometeorological stands were erected to record net radiation through net radiometer, soil temperature through soil heat flux plates and soil thermometers fixed at 5, 10 and 20 cms depth, wind speed and direction by using relevant sensors. Thermocouple wires were used wherever temperature accuracy is required up to 0.10 °C in addition to automatic sensors.

The canopy and canopy air temperatures were recorded with the aid of Instatherm Infrared thermometer.

The methodology adopted for the inference of the results obtained in the investigation were according to standard techniques. The crop growth parameters, yield components and yield were subjected to 5 × 4 factorial analysis of variance. The pooled analysis of variance for genotypic adaptation and yield performance over seasons as measure of stability were calculated. Matrixes for correlation coefficients and path analysis for different parameters to have an insight on the likely relationship among themselves was attempted. The regression models through step down approach were developed in accordance with the procedures to come up with the probable meteorological parameters as prime predictor for the expected production potential of soybean genotypes. The soil temperature modified by the canopy cover by different soybean genotypes on its amplitude and acrophase at different times of the day and night and other micrometeorological parameters were calculated.

2.8.4   Results

The results established the sensitivity of soybean genotypes to varying macro and micro meteorological factors. The weather elements had dictum from the very germinability of seed sown to the very termination of crop life cycle at maturity. The genotypic potential of seedling emergence was considerably influenced by the environmental variables in different growing seasons. Thus, the time needed for this phenophase varied from 6-8 days for 4 genotype MACS-201, MACS-58, PK-472 and MACS-13. The seedling emergence was early for all these genotypes is 5-7 days in Kharif. The genotype MACS-330 had an early seedling emergence within 4-6 days in all the seasons probably due to the rapid developmental rate of the coleoptile than the other genotypes. It was inferred that the soil temperature could possibly be an important microclimatic element to which the sensitivity of soybean seedling emergence was genotypic dependant.

The complex interaction of weather elements influencing soybean genotypes had a pronounced effect even on the inherent longevity of the crop and also different phenophases from seedling to senescence. These inherent genotypic differences were also altered by different sowing dates within each season eventually due to the environmental variation. The generative phase of soybean genotypes commenced late by delaying the sowing time within 20 days interval beginning from 15th October until 14th December during the rabi season in 1996 as well as in 1997. On the other hand, the other phenophases during reproductive and maturity time tended to have attained earlier by late sowing. In summer, the crop sown on 25th January commenced the vegetative growth earlier but needed relatively more time for other phenophases in later dates. This trend was also similar with extended sowing dates from 25th June in the kharif season.

The five soybean genotypes tested had diverse morphological growth features and yield potential. The genotype MACS-201 was of long duration. Its maturity period was influenced by the interaction of macro and micro meteorological elements. It matured in 100 days in Kharif, 110 days in summer and 120 days in rabi season. Its stature was also technically tall all through the crop growth period with profuse branching, more number of leaves, canopy dry weight and leaf area per square meter as well as the leaf area index. It was also superior to produce more number of pods per plant, number of seeds per pod, 100 seed weight and seed yield per hectare. The performance of 2 genotypes MACS-58 and PK-472 was not on par with this genotype with reference to the plant height, number of branches, number of leaves, leaf area per square meter, leaf area index, canopy dry weight, yield components and yield. These genotypes matured early in 104-110 days in summer, 107-114 days in rabi and 97 to 102 days in kharif. The genotype MACS-13 was significantly inferior in crop growth, yield components and yield, while it matured early in every season. The genotype MACS-330 was extremely short in height with poor branching and canopy profile. It developed less number of pods per plant and number of seeds per pod with low 100 seed weight. Its yield was extremely low. It produced 38.7 percent less yield than MACS-201 during rabi seasons, 40.90 percent in summer and 38.80 percent in kharif. The yield components and yield of all the five genotypes were severely reduced in summer than in kharif or rabi seasons. The ideal time to sow the crop was 25th June in Kharif, 15th October in Rabi and 25th January in summer for specific agro ecological situations of the experimental site. The crop attained luxuriant vegetative growth with maximum development of pods per plant, number of seeds per pod and 100 seed weight. Eventually, maximum production levels were realized from crops sown on these dates. The crops sown later on 14th December in Rabi season produced 23.21 percent less yield than the optimum date of 15th October in both the years. The yield reduction was to the extent of 25.80 percent by delayed sowings of soybean genotypes on 6th March than 25th January in summer. The magnitude of yield loss was similarly detected for the late sown crop on 4th August was 22.08 percent in the kharif season compared to the favorable time of seeding on 25th June.

The pooled analysis of variance showed that the soybean genotypes with stable yield performance under varying environmental conditions ranked in the order of MACS-201 < MACS-58 < PK-472 < MACS-13. The genotype MACS-330 produced less yield than the mean. Although it was also stable in performance with no significant deviation from zero its adaptability as a sole crop will not be possible for the peasants of Andhra Pradesh owing to its low yield. It may fit well as an intercrop.

The correlation coefficient analysis indicated that the plant height, number of leaves, leaf area per square meter, leaf area index, canopy dry weight, number of pods per plant, number of seeds per pod, 100 seed weight and the harvest index were all significantly interdependent which in turn conjunctively influence the economic yield potential. These associations with significant and positive correlations were recorded for every genotypes except number of seeds per pod with yield only in MACS-201. The path coefficient analysis which partitions the direct and indirect causes and effects on yield reveal the importance of canopy dry weight. This parameter was the common source for yield improvement of MACS-201, MACS-58, PK-472 and MACS-13 both directly as well as indirectly by the influence of all other morpho physiological parameters. Its indirect influences were more pronounced than its direct influence in promoting the yield of MACS-330. This underlines the need for attention on all physiological parameters that are sensitive to environmental variation in this genotype while a cursory attention only on canopy dry weight for its relative change in relation to environmental variation would suffice for proper yield variations.

The influence of macro meteorological parameters recorded substantial differences to modified responses of soybean genotypes. The maximum temperature recorded although the crop growth period during vegetative, reproductive and maturity phases had a significant and negative correlation on seed yield of all the five genotypes. The minimum temperature was of little importance. Its negative influence was significant only at maturity phase in MACS-201, MACS-58, PK-472 and MACS-13. The genotype MACS-330 was independent of considerable influence by the minimum temperature. On the other hand, morning and evening humidity had a pronounced influence all through the crop growth at vegetative, reproductive or maturity phase to impart significant positive correlation to seed yield in MACS-201, MACS-58, PK-472 and MACS-13. The prevailing humidity during vegetative phase, morning humidity during maturity while both morning and evening relative humidity during reproductive phase established positive and significant correlation with yield in MACS-330. The sunshine hours were significantly correlated with seed yield of all the five soybean genotypes. The association was negative for sunshine hours prevailed during vegetative, reproductive or maturity phase. The mean temperature during reproductive and maturity phases exerted significant and negative influence on seed yield of all the genotypes except MACS-13 in which this effect was significant for mean temperature in the reproductive phase alone. The growing degree days during reproductive and maturity phase exercised negative and highly significant association with seed yield of every genotype. The heliothermal units in the vegetative, reproductive as well as maturity phases had positive influence on yield. the correlation coefficients were highly significant.

The Absorbed Photo synthetically Active Radiation was best utilized to trigger the yield of soybean genotypes in rabi than kharif. The low absorption of Photosynthetically Active Radiation and its utilization reduced the yield of soybean genotypes exposed to climatic variations of summer. This micrometeorological parameter was also best utilized as indicated by better Radiation Use Efficiency in enhancing the seed yield of soybean sown on identified optimum than earlier or later dates within each season.

Multiple regresson equations were fitted for different crop growth parameters and seed yield with macrometeorological elements by step down regression approach. The yield of soybean genotype MACS-201 was described as a function of relative humidity in the evening during vegetative phase and heliothermal units in the reproductive phase and maximum temperature in maturity phase. The yield response of MACS-58 was best predicted through the fitted equation with estimated coefficients for morning relative humidity and mean temperature in the vegetative phase along with the heliothermal units in immaturity phase. Only one weather parameter i.e., maximum temperature in the maturity phase was adequate to predict the yield response of PK-472. The yield prediction of MACS-13 as governed by estimated coefficient for minimum temperature, morning relative humidity and heliothermal units in the vegetative phase, morning relative humidity and heliothermal units in reproductive phase and maximum temperature and heliothermal units in the maturity phase. The best fitted models for crop weather relationships to estimate the productivity level of MACS-330 was attained through the estimated coefficients of evening relative humidity and heliothermal units immaturity phase. It was inferred that the genotype specific models of soybean developed are to be further validated for wide application at different locations in semi arid regions.

2.8.5 Conclusions

The macrometeorological and micrometeorological studies on the performance of soybean and its probable introduction in the semi arid zone of Andhra Pradesh was attempted through extensive research at the main campus of Acharya N. G. Ranga Agricultural University during 1996-1998. The agro ecological semi arid zone of Andhra Pradesh was found to have a favorable environment for the protein and oil rich soybean to be nurtured in alfisols throughout the year in Kharif, rabi and summer seasons.

Its yield performance was outstanding in the rabi environment and was also excellent to be grown even in the kharif season. The dependable production of soybean seed yield in summer season despite low productivity than in rabi or kharif is still an admirable venture owing to its competitive price and consumer preference with several other traditional crops. The ideal sowing time was 15 October in the rabi season, 25th June in Kharif and 25th January in summer season to realise maximum seed yield. The choice of genotypes to be adapted is highly illusionary for MACS-201. In the event of non availability of the seed of this genotype the next options would be MACS-58 and PK-472. The low yielding genotypes MACS-13 and MACS-330 are suitable for diversification of less remunerative cropping systems owing to their short duration rather than to adopt them as a full season single crop.

Prediction models developed for probable estimates of yield for five genotypes serve as a milestone in adaptability of the genotypes suitable for new areas of introduction through the climatological data based approach.

The dreams of Mahatma Gandhi - The Father of the Nation in as early as 1930 to popularize soybean for its qualities and alleviate protein deficiency among Indian masses proved fruitful through the investigation.

3

Weather – Disasters – Management

Above it is full of me, below it is full of me, in the middle it is full of me.
I am in all beings, and all beings are in me. Om Tat Sat, I am it. I am
existence above mind. I am the one sprit of the universe. I am neither
pleasure nor pain. The body drinks, eats, and so on. I am not the body. I
am not the mind. I am He. I am the witness. I look on. When health
comes I am the witness. When disease comes I am the witness. I am
Existence, Knowledge, Bliss. I am the essence and nectar of knowledge.
Through eternity I change not. I am calm, resplendent, and unchanging.

- Isopanishad

3.1   An Overview of Disasters

Disaster is defined as "An event either man made or natural, sudden or
progressive, the impact of which is such that the social structure of the
affected community gets disrupted and fulfillment of all or some of the
essential factors of society is prevented".

Hazard × Vulnerability = Disaster

The super cyclone in 1999, drought in 2002 and 2009, extreme events
of temperature, floods and flash floods are the disasters that India faced in
the last decade. The consecutive flash floods over three major metro
cities in the same year 2005, Mumbai in July, Chennai in October and
December, Bangalore in October caused heavy damages to economy and
loss of life. There is a need to address these issues from technological,
social and economic points to bring sustainable alternate livelihoods and
to enhance resilience. During 1963-2002 the major natural disasters that
occurred around world are floods (32%), tropical cyclones (30%),
droughts (22%), earthquakes (10%) and others (6%).

61

3.1.1   Major Types of Disasters

- Hydrometeorological/Natural disasters: Climatic hazards and natural calamities like earth quake, famine, flood, cyclone, drought, dust storm, landslide, volcano and avalanche
- Manmade Disasters: Technological disasters like dam failure, nuclear accident, oil spill, gas leakage, fire, social disasters like riots and mass migration
- Localized (fire accident) and widespread (volcanic eruption)
- Major (earth quake) and minor (gas leak)
- Predictable (drought) and unpredictable (lightning and earthquake).

3.1.2   Disasters based on "Set on Time"

- Slow on set disasters which are predictable: Drought, famine and food shortage
- Quick on set disasters which are predictable: Cyclones, floods and typhoons
- Quick on set disasters which are unpredictable: Earthquake, landslides and avalanches.

3.1.3   Disasters based on "Response Time"

- Long response time: Drought and famine
- Short response time: Cyclones and floods
- No response time: Earth quakes and landslides.

D's of Disasters

Death, Disease, Damage, Destruction, Dislocation, Disruption.

Elements at risk

People, livestock, rural housing, houses vulnerable, crops, trees, telephones, electric poles, boats, looms, working implements, personal property, electricity, water, food supplies and infrastructure support.

3.1.4   Impact of Disasters on People

- Psychological manifestations
- Loss of property
- Loss of lives of near and dear
- Agony and sufferance undergone during the incident (children and women are most sufferers).

3.1.5   Scale of Disaster

This depends on

- Lead time available
- Intensity of hazard
- Duration
- Spatial extent
- Density of population and extent
- Time of occurrence
- Vulnerabilities existing in the elements at risk.

3.1.6   Assessment of Agricultural Losses due to Disasters

- Gross cropped area and affected area
- Areas completely damaged and partially damaged
- Damage to natural resources
- Crop yields under normal and partially damaged conditions
- Loss of potential output
- Loss of inputs
- Stage of crop at the time of disaster
- Prices used to calculate monetary value
- Agency making the assessment of damages and its purpose.

3.1.7   Disaster Management

The most important aspects of effective disaster management are precisely those activities that are undertaken and prepared at the times when there is no disaster.

Phases of Disaster Management

For administrative and policy convenience three distinct phases are identified for disaster management. These are:

(a) Disaster preparedness (prior to disaster)

(b) Rescue/relief operations (during disaster)

(c) Rehabilitation (after disaster).

3.1.8   Disaster Management Vision

(a) Mitigation: Hazard evaluation, risk assessment, preventive strategy, emergency planning and responding levels

(b) Preparedness: Regular observations, regular reports, detection (radars and satellite pictures) area mapping (topology and low lying areas)

(c) Recovery: Rescue, rehabilitation and reconstruction.

3.1.9 Community based Disaster Management

The top down approach failed to address specific local needs of vulnerable communities. Then, community based disaster management was recognized as an effective alternative for managing and reducing risks.

Objectives

- To prevent avoidable loss of life
- To minimize human sufferings
- To alert and inform public and authorities of impending risks and encourage activities to mitigate the same
- To minimize property damage and economic loss
- To speed up recovery and rehabilitation towards development.

Benefits

- Encourages high local people participation
- Encourages local traditional skills
- Cost effective because local resources can be used
- Enables community for self reliant
- Clears unrealistic myths
- Avoids panic situation
- Increased sense of ownership and collective interest.

The major hydrometeorological disasters that impact agriculture are cyclones, floods and drought.

3.2 Cyclones

Cyclone/cyclonic storm is defined as "The atmospheric disturbance which involve a closed circulation around a low pressure center, anticlockwise in the northern hemisphere and clockwise in the southern hemisphere". Also, "An intense vortex or a whirl in the atmosphere with very strong winds circulating around it in anti-clock wise direction in northern hemisphere and clock-wise in the southern hemisphere".

Cyclones are accompanied by heavy rains and are regular feature in coastal areas.

Tropical cyclones are most destructive natural hazards. The cyclones cause considerable human suffering in about 70 countries around the world. An average of 80 tropical cyclones form annually over the tropical oceans, of which, 30 occur in typhoon region of the western north pacific. The average for the Bay of Bengal and Arabian Sea is five. Some of the most destructive of these storms have occurred in this region (the severe tropical cyclone in Bangladesh in 1970 claimed three lakh lives). EI-Nino is generally associated with worldwide anomalies in the patterns of precipitation and temperatures as well as with patterns of tropical storms and hurricane activity, the behaviour of subtropical jet streams and many other general circulation features over various parts of the world. The magnitude of hurricanes is assessed with the "saffir-simpson scale" which takes into account maximum sustained winds and minimum storm pressure. Hurricane Georges (September 1998) and Hurricane Mitch (October 1998) are the most devastating among hurricanes of all times. The hurricane Mitch is category 5 hurricane and one of the most powerful Atlantic hurricanes, with 290 km/hr winds and a minimum storm pressure of 906 hPa and had quite a long life span (14.5 days) which explains why it turned out to be the deadliest of the century. It caused loss of life, destruction of property, damage to food production, food reserves and transportation systems as well as increased health risks. In May 2002, Cyclone Kesiny hit Madagascar affecting more than half a million people, making them homeless or in need of emergency food, shelter and drinking water. Up to 75% of the crops were destroyed. The cyclones on 17-18 October and again on 29-30 October 1999 in Odisha, India caused devastating damage. The cyclones on 29-30 October with wind speeds of 270-300 km/h for 36 hours were accompanied by torrential rain ranging from 400 to 867 mm over a period of 3 days. The two cyclones together severely affected around 19 million people in 12 districts. Sea waves reaching 7 metres rushed 15 km inland. About 2.5 million livestock perished and a total of 2.1 million hectares of agricultural land was affected.

3.2.1 Formation and Movement of Cyclones

According to wave theory, cyclones form at a wave like twist or perturbation on the front. The waves thus formed may be of two types viz., stable and unstable. The stable waves form and dissipate without any visible effect on the weather. In case of unstable waves, they grow in amplitude and go through a cycle. The unstable waves require a wavelength varying from 500 to 3000 km. Thus, the cyclogenesis (cyclone formation) occurs where a frontal surface is distorted into a wave shaped discontinuity.

- Tropical cyclones are the off spring of "ocean – atmosphere" interactions

- These are powered by heat from the sea, driven by easterly trades and temperate westerlies, the high planetary winds and their fierce energy.

There are marked seasonal variations in their place of origin, tracks, and intensities. For example, in Bay of Bengal cyclones occur during Pre-monsoon (April-May) and post monsoon (October–December) and in Arabian Sea in May and from October to December. These behaviors help in predicting the movements of cyclones.

3.2.2   Life Cycle of a Tropical Cyclone

(a) *Formative stage*: The system grows to a cyclone from a trough of low, low pressure area, depression and after about a week at the end of this stage eye and wall cloud will be formed. Pressure difference between center of tropical cyclone and the outer most isobar is approximately 10 hPa

(b) *Immature stage*: Central pressure falls and wind speed increases. Duration is 12-24 hours. Lowest pressure and highest wind speed develops at this stage

(c) *Mature stage*: No further fall of pressure system. Strengthening of wind takes place and system expands in size. Symmetry may be lost

(d) *Decay stage*: System weakens rapidly to deep depression, depression, low pressure area when it enters land area.

3.2.3   Structure of Cyclone

- A full grown cyclone is a violent whirl in the atmosphere with 150-1000 km across and 10-15 km high

- Cyclones are intense low pressure areas from the centre of which pressure increases outwards

- A gale winds of 150-250 kmph or more spiral around the center of very intense system with 30 to 100 hPa below the normal sea level pressure.

A well developed cyclone consists of:

(a) *Eye*: The central calm region of the storm is called "Eye". The shape may be circular or elliptical. The diameter of the eye varies between 10-15 km and is a region free of clouds and has light winds, clear sky, lowest pressure, warmest temperature and descending motion

(b) *Wall Cloud Region*: Around the calm and clear eye, there is the wall cloud region of the storm which is of about 50 km in extent. Gale winds, thick clouds with torrential rain prevail accompanied by thunder and lightening

(c) *Outer Periphery*: Away from wall cloud region, wind speed gradually decreases. However, in very severe cyclonic storms or super cyclones, wind speeds of 50-60 kmph can occur even at a distance of 600 km from the storm center

(d) The average movement is 400 km/day. Wind speed 200 km/hr. Rainfall 50 cm/day. The storm surges as high as 3 to 12 meters inundate inland up to 30 km approximately. Very heavy rainfall of the order of 30 to 40 cm occur in respect of severe cyclones and of the order of 20 to 30 cm in case of cyclones.

The severity/destruction due to cyclones depends on:

- The amount of the pressure drop in the center and the rate at which it increases outwards gives the intensity of the cyclones and strength of winds

- The storm surge which depends on cyclone intensity, bathymetry of the coastline, coastal configuration, angle at which the cyclone strikes the coast and time of landfall

- The amount of rain

- The characteristics of gales.

3.2.4 Classification of Cyclones

The criteria followed by IMD (India Meteorological Department) to classify the low pressure systems is as follows:

System	Pressure deficiency @ (hPa)	Associated wind knots (kmph)
Low pressure Area	1.0	<17(32)
Depression	1.0 -3.0	17-27(32-50)
Deep Depression (DD)	3.0 -4.5	28-33 (51-59)

Table contd...

System	Pressure deficiency @ (hPa)	Associated wind knots (kmph)
Cyclonic Storm (CS)	4.5-8.5	34-47(60-90)
Severe Cyclonic Storm	8.5-15.5	48-63(90-119)
Very Severe Cyclonic Storm (VSCS)	15.5-65.6	64-119(119-220)
Super Cyclone	>65.6	>119(>220)

@ system's centre to outer edge (periphery)

3.2.5 Effects of Cyclones

The cyclones have relatively large scale effects and affect about a third of the world population. Damage is caused not only by immediate effects of the wind, but also to a high degree is induced by the effects of storm surges and ocean waves directly generated by tropical cyclones. The tropical cyclones cause irreparable damage.

3.2.5.1 *Damage to General Systems*

- The principal causes of destruction by topical storms are fierce winds which cause havoc and torrential rains with associated flooding and high storm tides (the combined effect of storm surge and astronomical tides) resulting in salinity
- Most casualties are caused by coastal inundation due to storm tides. Casualties also occur due to drowning, injuries, non availability of transportation at right time, deterioration of health conditions and improper sanitation
- Collapse of buildings, falling of trees, flying debris, electrocution and diseases from contaminated food and water in the post cyclone period contribute alarmingly to loss of life and destruction of property
- The indirect damages due to cyclones are caused by viral diseases, respiratory infections, snake bites, skin infections, conjunctivitis and starvation due to food shortage
- The heavy rain causes destruction of vegetation, crops, livestock, contamination of water supply, land subsidence and flooding of inland areas
- Electric poles get uprooted and damaged which results in loss of electric power and disruption to communication systems

- The economic and social impact depends on timing of cyclone, duration, velocity and extent of area.

3.2.5.2 *Damage to Agricultural Systems*

- The loss to agriculture from tropical cyclones can be due to direct destruction of vegetation, crops, orchards and live stock due to heavy rains
- The wind damage through airborne sea salt, which occurs within a few hundred metres of the coast that spray salt on coastal areas, make it impossible to grow crops sensitive to excessive salt
- Long term losses of soil fertility from saline deposits also occur over land flooded by sea water
- Small scale fisheries are also hit by cyclones
- The rise of sea level would adversely affect the coastal ecosystem, by causing damage to irrigation infrastructure such as canals, wells and tanks
- Disruption of the agricultural transportation system
- Loss of a portion of the future harvests due to the destruction of standing crops
- Damage to agricultural and animal property, off-shore and on-shore installations
- Loss in agricultural productivity due to disruption of the work force and to other activities
- Typhoons have been known to implicit severe damage on agriculture. A few examples are:
  - In southern Hainan on 2nd October 1999, timber trees and rubber trees together 25 million were damaged
  - A typhoon that struck Thailand on 4th November 1989 wiped out 1,50,000 hectares of rubber, coconut and oil plantations and other crops.

3.2.6 Beneficial/Positive Effects of Cyclones

- Not all the impacts of cyclones are negative. Some reports cite beneficial affects of tropical cyclones
- The areas which are in the periphery of cyclones enjoy the benefit of rainfall

- It was observed that increased water availability in water critical regions makes agricultural production less susceptible to the dry period

- It was estimated that nine major hurricanes in the United States since 1932 terminated dry conditions over an area of about 622,000 km^2.

3.2.7 Management of Agricultural Systems affected by Cyclones

Disaster preparedness for impending cyclones refers to the plan of action needed to minimize loss to human lives, damage to property and agriculture. The effectiveness of disaster preparedness ultimately depends on the effectiveness of planning and response at a district or local government level. Preparedness for cyclones in agricultural system include:

- Early harvesting of crops at physiological maturity

- Safe storage of harvest

- Cultivation of short duration crops or varieties which are not easy grain shedders

- Spreading type of crops or varieties may be preferred

- Alternative cropping pattern in areas affected by salinity needs to be implemented

- Reclamation and treatment of soils

- Maintenance of ecological balance through large scale plantation

- Wage employment under various rural development schemes

- Activities in farm sector may also include growing vegetables, nurseries and kitchen gardens

- Draught animals and dairy cattle to be provided to farmers

- Arrangements for restoring insemination centers for local breeds of cattle have to be made

- Activities in the non-farm sector for engaging more people in occupations such as fishing, fish drying, pig rearing, coir production, rearing of cows, buffaloes, bulls, bamboo production, lime making, poultry, goat rearing, boat building, mat weaving, electrician work, blacksmith work, toy making, pottery and petty trading

- Restoring the traditional livelihoods of the affected populations through the provision of food/cash for work
- Irrigation canals and embankment of rivers in the risk zone should be repaired to avoid breaching
- As the storm approaches the area, secure safety for as much of the property as possible
- Facilitating work for community based rehabilitation and restoration activities, such as reclamation of agricultural land
- Providing a model for food-for-work programme that is capable of reaching the most vulnerable and marginalized, so that other large-scale programmes can integrate vulnerability and equity perspectives.

3.2.8   Protection of Animals during Cyclones

- Fodder and water shall be kept in enough quantities
- Proper treatment to sick animals
- Vaccination
- Burry the dead animals immediately
- Use proper medicines to external parasites
- Treat water with bleaching powder
- Procure medicines for post cyclone situations
- Allow the animals to stay in thatched/leafy sheds
- Insure the animals.

3.2.9   Cyclone Mitigation Measures in Agricultural Systems

- Hazard prone areas should be subject to land reform
- Afforestation should be encouraged
- Risk zone mapping and analysis of land use patterns should be undertaken
- Land use legislation and building regulations should be established
- Raised platforms for livestock, emergency food, grain storage facilities, drinking water storage and wells with covers should be built.

3.2.10 Advisory for Hudhud Cyclone affected Districts in Andhra Pradesh & Odisha

Cyclone 'Hudhud' made a landfall on 12th October 2014 at Visakhapatnam and caused serious damage to agriculture in four districts each in Andhra Pradesh and Odisha. State wise crop advisory is given below for reference and adoption and contingency measures to minimize and prevent further damage in standing corps.

Andhra Pradesh

North coastal districts of Andhra Pradesh (Visakhapatnam, Srikakulam, Vizianagaram and East Godavari) suffered widespread damage due to gale accompanied by downpour.

Crops affected	Damage	Advisory
Rice	Crop in various growth stages affected due to partial or complete lodging due to high speed winds and partial inundation due to accompanying rain. Long duration crop varieties are in panicle initiation to grain filling stage while medium maturing varieties are in flowering stage. Rice at flowering stage has been seriously affected. Flood is imminent in low lying villages of Srikakulam district in the command areas of Nagavali river due to heavy inflows.	• Drain out excess water by making alleys at periodic intervals in the lodged crop • Take up staking of plants at grain filling stage • After flood water is receded, apply 25 kg urea and 10-15 kg of MOP as booster dose to long duration varieties • Apply 15-20 kg of Potash or spray of multi-K (13-0-45) @ 10 g/l of water for medium duration varieties • Early planted rice which is in maturity stage may be sprayed with 5% salt solution to prevent seed germination • Spray hexaconozole @ 2ml/l as prophylactic spray to prevent occurrence of sheath blight/blast disease incidence • Spray streptocycline @ 0.1 gm/l if bacterial leaf blight is noticed (varieties: BPT – 5204, MTU-1001, MTU-1075) • In partially lodged crop in panicle initiation stage especially in long duration and susceptible fine rice varieties, keep a watch on brown plant hopper population at the base of plants. Observe if population is in rising trend as conditions are conducive for pest outbreak if drizzly and cloudy weather conditions prevail for few more days. Apply need based spray immediately after cessation of rain based on close monitoring of pest incidence.

Table Contd...

Crops affected	Damage	Advisory
Maize	Lodging and stress due to excess soil moisture	• Staking of lodged plants • Provide quick drainage of excess water by opening furrow • Harvest cobs at physiological maturity or marketable green cobs or for fodder purpose in case crop is badly damaged • Undertake earthing up of plants in partially affected fields • In mild to moderately affected fields due to lodging/uprooting, take up prophylactic or need based spray to prevent fungal diseases (blight & rot).
Oilseeds (groundnut and sesame)	Lodging in sesame, water logging in groundnut	• Harvest at physiological maturity of pods in sesame, undertake quick drying, threshing, safe storage/marketing • Provide quick reduction in soil moisture in groundnut fields to prevent premature germination by opening of drainage channel immediately after flood water recedes and harvest at physiological maturity • In slightly affected water logging groundnut fields take up need based spray of fungicide (chlorothalonil @ 2g/l) to prevent late leaf spot incidence • In case of severe shortage of green fodder to milch cattle, harvest groundnut and use haulms for feeding.
Short duration pulses and pigeon-pea	Water logging in pulses and lodging in pigeon pea	• Harvest late planted pulses (blackgram & greengram) at physiological maturity • Provide early drainage from water logged fields • Use damaged plants as fodder to milch cattle • Stake lodged plants in pigeonpea and undertake earthing up of plants at the earliest opportunity for tillage • In badly affected fields due to water logging, harvest crop for fodder purpose.

Table Contd...

Crops affected	Damage	Advisory
Sugarcane	Lodging and water logging	• Staking of lodged plants • Provide quick drainage by opening furrows • Earthing up of affected plants.
Vegetable crops	Lodging and water logging in tomato, brinjal, raddish and cucurbit crops	• Standing crops damaged due to lodging and water logging • Harvest produce at the earliest opportunity. Undertake shifting, grading and marketing of produce • Harvest at physiological maturity • Undertake nipping of apical buds to induce sympodial branching to compensate for production loss • Apply light booster dose of fertilizer.
		• If soil application is not possible, apply foliar spray of 2% urea/DAP 0.1% MOP • To prevent further premature drop of flowers/fruits, apply NAA hormone (40 ppm) • Take up community nursery to supply seedlings for cultivation of vegetable crops in post flood situation in badly affected villages To prevent root, foliar and fruit diseases (rot, leaf spot, blight), apply foliar spray of fungicides/bactericides (Mancozeb 2g/l, carbendazim, 1g/l, copper oxy chloride 3g/l, plantamycin 0.6g/l) or soil drenching with copper oxy chloride 30g/l).
Coconut, Banana, Cashew & pulpwood plantations (Casurina, Eucalyptus etc)	Breaking of branches, lodging, partial or complete uprooting of trees	• Provide early drainage of excess water from orchards/plantation • Propping/staking of partially lodged/uprooted trees • In banana, remove lodged plant to allow one good sucker to replace the lost plant • Prune broken branches of trees (flat cut at breakage point) and apply Bordeaux paste to cut end

Table Contd...

		• Apply booster dose of fertilizer after optimum moisture conditions return in the orchard • Apply need based plant protection measures as there is likelihood of disease occurrence in stressed plants • For coconut, apply copper oxychloride 3g/l of water in whorls after clearing the broken/dropped leaves in young trees • Fresh planting may be taken up in place of the lodged, uprooted older trees • Under take gap filling in the orchard in case of complete loss of plants due to uprooting in plantation crops.

Odisha

Gajapati, Koraput, Malkangiri and Rayagada were the worst affected districts. Details of crop damage and contingency measures are given below:

Crops affected	Damage	Advisory
Paddy	Lodging, partial submergence	• Staking of lodged plants • Make alleys at intervals to clear excess water • Apply N @ 20 kg/ha for quick recovery in late rice or spray 2% urea to crop in flowering stage • Heavy rainfall is likely to trigger outbreak of swarming caterpillar and cut worm (panicle stage) in late rice. Apply need based spray of chlorpyriphos @2 ml per 1itre • Prophylactic spray of copper oxychloride, streptocycline and imidacloprid to prevent bacterial leaf blight, blast and insect attack (BPH) where crop is badly affected and loss is complete, contingency crops like short duration greengram followed by sesame are suggested.
Pulses and Groundnut	Water logging	• Provide quick drainage • Harvest at physiological maturity to prevent premature germination • Use residue for fodder purpose.
Mango and Papaya, Teak and Cashew	Lodging, breaking, uprooting of plants	• Provide quick drainage in orchards/plantations • Prune broken branches in orchard crops and apply Bordeaux paste to cut ends

Table Contd...

		• Propping of papaya plants and harvesting of marketable fruits
		• Earthing and staking of lodged plants and application of the booster dose of fertilizer.
Vegetable crops	Tomato, Brinjal, Chillies	• Provide quick drainage
		• Practice earthing up of plants
		• Harvest at physiological maturity/marketable produce
		• Take up plant protection measures through foliar spray of fungicides (copper oxy chloride 3g/l) or drench soil at the base of plant to prevent rot/wilt
		• Apply light booster dose of fertilizer under optimum soil moisture conditions to stimulate growth
		• Practice nipping of apical buds to promote sympodial branching.

3.3   Floods

Flood is defined as "Any relatively high water flow that overtops the natural or artificial banks in any portion of a river or stream". Also, "Flood is an extreme event of hydrological cycle, which occur due to the excess run off after meeting the soil moisture recharge, percolation, evaporation and storage".

When a bank is overtopped, the water spreads over the flood plain and generally becomes a hazard to society. Floods are caused by severe thunderstorm or prolonged heavy (including monsoonal) rainfall. These floods occur mainly in areas where the soil is not well drained or are clayey and hence non-drainable.

Floods are among the greatest natural disasters known to mankind. According to historical record, 1092 flood disaster events occurred since 206 BC in China during a period of 2155 years, averaging once in every two years. In China, about 8% of the land area is located in the mid and downstream parts of the seven major rivers in the eastern and southern parts of the country that are prone to floods. About 50% of the total population of the country lives in these areas. The number of people affected by floods in the world during 1991 to 2000 was around 1.5 billion. The frequency of floods had increased in the Odisha state of India. Between 1834 and 1926, the state experienced floods once in four years, which rose to once in two years after 1926. The state experienced nine bouts of floods within a span of 15 days in 2001, an all time high, damaging 2.12 million hectares of standing crop.

Flood plains are the areas of highest productivity. Some of the most flourishing ancient civilizations were in the flood plains as proximity to rivers enabled them to enhance agricultural productivity. In many areas, annual floods are welcomed by the populations settled in such areas since the increased availability of water resources could be harnessed for greater agricultural output. But, such activities are not without risk since they could further increase the risk of flooding hazards.

Floods are a function of the climate (variability in rainfall pattern, occurrence of storms) as well as hydrology (shape of river beds, intensity of drainage, and debit flow of rivers) and soil characteristics (moisture absorption capacity). Flooding occurs primarily when water due to rain from various types of weather phenomena or snowmelt accumulates faster than soils can absorb it or rivers can carry it away. The countries most affected by floods are Brazil (15%), USA (12%), Peru (11%).

3.3.1 Classification of Floods

Floods vary in degree of sensitivity in terms of, areas in extent and magnitude in depth, thus classified as:

(a) Major and

(b) Minor floods

(a) *Minor floods:* Inundation may or may not be due to over flowing of water on the banks of the water bodies. Flooding may be due to the accumulation of excessive surface runoff in low lying flat areas or topographically depressed terrains

(b) *Major floods*: These are caused by overflowing of rivers and lakes, serious breaks in dikes, levees, dams and other protective structures by uncontrollable releases of impounded water in reservoirs and by the accumulation of excessive runoff. Flood waters cover much larger areas and spread rapidly to adjoining areas of relatively lower in elevation.

3.3.2 Types of Floods

(a) Coastal flooding

(b) River flooding

(c) Flash flooding

(d) Urban flooding

(e) Groundwater flooding

(f) Sewer flooding.

3.3.3   The Causes of Floods

Floods are caused by excess rainfall in river catchment areas. The main causes of floods are:

- Natural weather events such as cloud burst or heavy thunderstorms that produce heavy to very heavy rainfall over a very short period of time
- Prolonged and extensive rainfall over urban areas
- Astronomical tides combined with the storm surges
- Lack of maintenance of infrastructure of existing water bodies
- Insufficient drainage networks
- Lack of mechanism to divert excess rain water in to drainage systems
- Poor maintenance of drainage systems
- Inappropriate development in flood plains
- Lack of good flood defense systems.

The damage / severity of floods depends on

- Extent of area effected
- Duration of flood
- Velocity of flood currents
- Types of crops grown in the area
- Timing of floods
- Sand casting
- Erosion
- Rise in water table.

3.3.4   The Negative Impacts of Floods on Agriculture

- Damages to irrigation systems like breaches in embankments
- Damages to drainage systems
- Lives and property of farmers at risk
- Over flow of rivers inundate vast areas of agricultural land and submerge crops
- Water logged conditions cause failure of crops

During the non-growing season (fallow season)

- Loss of top soil and soil nutrients

- Sediment transport, landslides, soil compactation, erosion and deposition of undesirable deposits and debris
- Permanent damage to perennial crops, trees, livestock, buildings and machine
- Debris flow, Anaerobic processes and cessation of farming in flood plains

During the growing season

- Contamination of water and water logging and lodging of standing crops
- Loss of soil nutrients, interruption to use of pastures and soil erosion
- Greater susceptibility to diseases and insects and grain spoilage because flood waters are favourable breeding places for insects and diseases
- Interruption to tillage, planting, crop management and harvesting.

3.3.5   The Positive Impacts of Floods on Agriculture

- Soil fertility improves due to silt deposition in flood plains
- Residual soil fertility after floods favour succeeding crops.

3.3.6   Flood Management

- Various devices may be employed in order to control an excess flow of water so that a flow may be prevented, or at least, the worst effect reduced. These devices include engineering works, embankments, detention reservoirs, the adaptation of river channels and facilities for flood diversion
- When the cause is excessive rainfall over the basin, then leaves and floodwalls along the river are used to confine the water to the river channel and thus accelerate the floodwater flow to the sea or storage reservoir
- When the cause of flooding includes tidal influence then control gates at the river mouth are used to regulate the tidal effect
- On farm storage in lowland rice fields
- Construction of a series of low earth dams
- Reforestation and terracing of hilly lands
- Building houses on mounds or higher ground
- Building a well defined warning system

- Control over the river and land through measures such as flood proofing, forecasting and warning
- Use of GIS as an effective tool for developing flood emergency response.

3.4   Droughts

The term "Drought" means scarcity of available water which is normally required for an activity (drinking, agriculture and industry) in a region or a place for sufficiently longer period. Drought is a slow creeping disaster and an advance signal of famine. Conventionally, for climatological purposes, rainy day has been defined as "A day with rainfall of ≥ 2.5 mm/day". A year is considered as a moderately drought year if the annual rainfall is less than 80 and above 50 percent of the normal rainfall and a year with rainfall less than half of the normal is considered as severely drought year. Each severe drought year was considered equivalent to two moderate drought years. Number of events of rainless period of at least 14 consecutive days during the monsoon (June to September) is called "Dry spell". Number of events when cumulative rainfall in three consecutive days is more than 100 mm is considered as extreme rainfall event.

3.4.1   The Causes of Drought

- Failure of monsoon, inadequate water conservation efforts resulting in lack of water supply, contamination of water, inadequate storage, insufficient conveyance facilities or abnormal demand
- Climatic fluctuations and variations in earth, ocean and atmospheric relationship in precipitation
- Human activities such as over grazing, poor cropping methods, improper soil conservation techniques and extreme deforestation.

3.4.2   The Severity of Drought

This depends on

- The fundamentals of the economy of the region
- The drought mitigation policy of the region.

3.4.3   Classification of Drought

- *Meteorological drought*: A situation in which the rainfall deviates apparently below normal for an extended period. The lack of water cause serious imbalance in the affected areas. As per precipitation, the below normal departure of rainfall was classified by the IMD, India, as mild (1-25%), moderate (26–50%) and severe (above 50%). This is also a period of abnormally dry weather which is as per spatial extent sufficiently prolonged. Droughts are considered as large scale (up to 25% area is affected) and worst (from 26 to 50% area affected).

- *Hydrological drought*: If meteorological drought prolongs it results in hydrological drought, which is marked by depletion of surface water and consequent drying of reservoirs, lakes, streams, rivers, cessation of surface runoff and also fall in ground water level. On account of urbanization and industrialization all over the world, hydrological drought claimed its immense importance.

- *Socio economic drought*: A situation where water scarcity adversely effect the economy of the region. The consequent effects are unemployment and migration of people for livelihood.

- *Agricultural drought*: It is related to soil moisture and occurs when available soil moisture is inadequate for healthy cop growth causing extreme stress and wilting. The soil moisture is so diminished that the vegetation can no longer absorb water from the soil rapidly enough to replace that lost to air by transpiration. This is also a situation involving a shortage of precipitation which adversely affects crop production.

Agricultural drought is again classified into five major types (in dryland areas):

- ❖ *Early season drought*:   The early season drought occurs due to delay in commencement of sowing rains. Farmers sow the seed by taking advantage of early rains. A long dry spell may lead to withering of seedlings and poor crop establishment.

- ❖ *Mid season drought*: This occurs in association with long gaps between two successive rain events, if moisture stored in the soil falls short of water requirement of crop during the dry period. On other occasions the mid season drought may be associated with low and inadequate rainfall in the growing season to meet the crop water needs as per the phenological stage.

- ❖ *Late season or terminal drought*: If the crop encounters moisture stress during the reproductive stage due to early cessation of rainy season and rise in temperature the situation hastens the process of maturity.

❖ *Apparent droughts*: The rainfall of a region may be sufficient to raise one crop of low water requirement (jowar), but, may not be enough to meet the water requirements of another high water requirement crop (paddy). Therefore, mismatching of crops in relation to water availability pattern leads to apparent drought.

❖ *Permanent drought:* This is associated with inadequacy of rainfall/ soil moisture to meet the water requirements of the crops during most of the years. These areas are called drought prone areas.

3.4.4 The Negative effects of Drought

- Droughts often stimulate sequence of actions and reactions leading to long term land degradation

- Clay soils shrink tremendously which is also known as subsidence

- The ground and the flora are so badly damaged that their resilience is impaired or even destroyed permanently resulting in famine and finally desertification.

- Trigger forest, bush and grass fires

- Conflicts may also occur among neighbouring states sharing common water resources

- Food shortages, speculation, hoarding, forced liquidation of livestock at depressed prices.

3.4.5 The Positive effects of Drought

- Although drought impacts are primarily negative, sometimes droughts do carry positive impacts

- Moderate droughts in the post flowering maturity stage of sugarcane helps to increase the sucrose content

- Mosquito reduction

- Reduced cost of snow removal in snowfall regions and other related activities

- Emergency water conservation leading to establishment and construction of permanent and efficient water storage systems.

3.4.6 Response of Plants to Drought Conditions

- The upper leaves of the plant maintain their physiological activity and the lower leaves are the first to wilt or dry up during a drought

- Under drought conditions plants develop short lateral shoots, tiny leaves from the axils of the dropped old leaves

- If the drought occurs during the seed formation the seeds become small and shrivilled

- The greatest yield reduction takes place when drought occurs during the beginning of the heading stage in paddy crop.

3.4.7 Mitigation Measures

Drought is a recurring phenomenon and its occurrence cannot be avoided. Therefore, there is a need to develop infrastructure for mitigation of drought. The impacts of drought can be minimized through the application of science and technology and certain plans. These are

- Deep ploughing, deep furrow, mulching and mixed cropping

- Soil moisture conservation through watershed development, check dams and contour bunding

- Developing drought tolerant varieties

- Inter bund treatments like key-line and appropriate tillage, summer tillage, ploughing/sowing across the slope

- Green manure treatment

- Vegetative barriers

- Ridge furrow configuration for planting and opening of conservation furrows

- Short duration crops

- Integrated watershed management for development of farming systems and best practices

- Regularization of runoff and storage in medium size reservoirs

- Rehabilitation of wastelands

- Space borne measurements for early warning

- Continuous monitoring of the environmental variables

- Reasonable buffer stock of food grain and fodder

- Deepening of old wells, digging of new wells and drilling of boreholes

- Construction of multipurpose dams

- The overall policy, legislation and specific mitigation strategies should be in place well before a drought

- Timely availability of credit, postponement of revenue collection and repayment of short term agricultural loans

- Implementation of crop and livestock insurance schemes
- Provisions for cattle camp in drought affected areas
- Early warning and drought monitoring on the basis of long, medium and short term forecasts
- Distribution of free food from the government stock/granaries
- Provision of employment to the poor, marginal agriculturists and landless labour
- Power at subsidized rates
- Alternate employment for people in government sponsored relief schemes
- Education and training to the people
- Participation in community programmes
- Bring public awareness of drought and water conservation
- Media awareness programme.

3.4.8 Drought relief Measures Relating to Livestock

- Preparing and implementing fodder movement plan and fodder contingency plan
- Fodder conservation techniques such as supply of chaff cutters
- Organization of cattle camps
- Discourage distress sale of animals
- Mobile health units for cattle
- Vaccination of animals
- Community fodder cultivation and planting of perennial trees for fodder
- Encourage backyard small poultry units
- Ensure supply of chicks/ feed to small poultry farmers
- Arrange soft loans.

3.5   Other Disasters, Risks and uncertainties that Effect Agriculture

These are temperature, wind, storms, (thunder, dust and hail storms) defective insolation and tornadoes.

3.5.1 Temperature

Temperature is essential for all plant physiological processes, gaseous exchange between plant and environment and stability of plant enzymatic reactions. However, both cold and heat waves and abnormal soil temperatures are adverse to crop growth and development.

Cold Waves

Wind chill factor is taken into consideration while declaring coldwave situation. The wind chill effective minimum temperature is defined as the effective minimum temperature due to wind flow, e.g., when the minimum temperature is 15 °C and the wind speed is 10 mph, then the wind chill is 10.5 °C. Departure of wind chill from normal minimum temperature is from – 5 °C to – 6 °C where normal minimum temperature is > 10 °C and from – 4 °C to 5 °C elsewhere, cold wave is declared. Wind chill is used to declare coldwave and when it is < 10 °C only. Also coldwave is declared when wind chill is < 0 °C irrespective of the normal minimum temperature for those stations. Departure of wind chill from normal minimum temperature is from – 5 °C to – 6 °C where normal minimum temperature is > 10 °C and from 4 °C to 5 °C elsewhere, it is declared as a severe coldwave.

- During winter (December - February) temperature decreases. This fall in temperature may cause damage to the crops. If the temperature drops on freezing or below, a frost may occur which causes severe damage to the crops. Threat of frost is danger to crops.

Frost

Frost is a form of condensation that forms on cold objects when the dew point is below freezing

Frosts are of two types:

- *Advection or airmass frost*: This results when the temperature at the surface in an airmass is below freezing
- *Radiation frost*: This occurs on clear nights with a temperature inversion.

Impact of advection Frost

There is a special case of frost caused by loss of heat by evaporation. This occurs when cold rain showers wet the leaves and are then followed by dry wind. This is advection frost. The usual effects of advection frost are:

- The injury and death caused by frost is due to the formation of ice crystals in and outside the plant cells

- During dormancy, plants can withstand lower temperatures upto $-20\,°C$
- Once growth has commenced, temperatures of few degrees below freezing point may be fatal
- The cell sap gets frozen below $0\,°C$, as also between cells
- Extra cellular ice formation occurs followed by withdrawal of water from the cell
- The protoplasm may become dehydrated and brittle resulting in mechanical damage or the cell may contract and damage the protoplasm.

Management of advection frost

- For protection of most field crops, the only satisfactory solution to the problem of advection freezing is to avoid it as far as possible by planting after the damage is past and by selecting varieties which will mature before the beginning of the risk

Impact of Radiation frost

- Arrange for smoking nearby the field to protect crops
- During evening hours wipe out the water droplets by pulling rope
- The damage due to radiation frost differs from the above freeze damage in degree and its spotly occurrence
- The radiation frost damage is critical during critical stages of growth
- Young seedlings may be killed
- Flowering stage is most prone
- Crops like potato, tomato and melons are vulnerable right up to maturity
- For most field crops and orchard crops flowering stage is most critical for frost damage
- Frosty nights followed by warm sunny days produce a sunclad on orchard fruits, considerably reducing their production.

Management of radiation frost

- The management of radiation frost can be grouped into (a) Passive methods and (b) Active methods.

Passive methods

- Clean cultivation
- Plant selection
- Maintenance of soil moisture
- Inter cropping
- Nutrient management
- Pest management
- Proper pruning
- Trunk painting
- Soil covers
- Wrapping plants with insulating material and enclosing the basal part of the plant
- Proper site selection
- Choice of growing season
- Breeding of cold resistant varieties

The above methods can be followed even for advection frost also. These passive methods do not involve any modification of environment.

Active methods

The active methods of frost protection are many, like use of

- Heaters
- Wind machines
- Sprinkling water
- Following weather forecast for better management of crops.
- Surface irrigation
- Furrow irrigation
- Foam insulation.

Fog

Fog is essentially a dense cloud of water droplets, that is close to ground. It forms when the difference between air temperature and dew point is generally less than 2.5 °C. The different types of fog are radiation fog, advection fog, evaporation fog, precipitation fog, upslope fog, valley fog, freezing fog, frozen fog, hail fog and smog.

Impact of Fog

Fog plays an important role not only in hydrological cycle but also on agriculture. Its impacts on land, sea and air transport is significant.

- Direct contact with acid fog water may harm vegetation and crops
- Fog plays significant role also in physical interactions within plant canopies thereby growth
- Reduction in PAR and cold stress are also major impacts of fog
- Fog provides congenial conditions for diseases and insect pest development.
- Smog is a trigger of pulmonary diseases and reduced day length can lead to depression

Management of fog

- Heating of fog layer (to evaporate droplets)
- Downwash mixing (to entrain drier air)
- Hydroscopic treatment (ice seeding) to precipitate out
- Use of fog breaks to prevent formation and movement into an area
- Mixing and evaporation promote improved visibility within an hour.

3.5.1.1 *Heat Waves*

In India, heat wave is defined as the condition over a prolonged period when the "departure of maximum temperature from normal is +4 °C to +5 °C or more for the regions where the normal maximum temperature is more than 40 °C and departure of maximum temperature from normal is +5 °C to +6 °C for regions where the maximum temperature is 40 °C or less". When actual maximum temperature remains 45 °C or more irrespective of normal maximum temperature, heat wave is declared. In addition, heat wave is declared, under special situations, when the maximum temperature of a station reaches atleast 40 °C for plains and atleast 30 °C for hilly regions.

When the "Departure of maximum temperature from normal is +6 °C or more for the regions where the normal maximum temperature is more than 40 °C and + 7 °C or more for regions where the normal maximum temperature is 40 °C or less" is defined as a severe heat wave condition.

- These are very harmful during summer. The harmful effects include shedding of fruits, plants and drying of water resources.
- Loss of water by evaporation from irrigation channels

- Transpiration increases from plants beyond recouping levels
- Plants tend to wilt and die owing to rapid desiccation
- Hot winds cause shrivelling effect at milk stage of all agricultural crops.

Management of heat waves

- Adoption of specific agronomic practices like shelter belts and choice of varieties
- Improving weather forewarding mechanisms
- Change of microclimate by means of cropping systems
- Translating weather information into operational management practices
- Crop insurance.

3.5.2 Wind

- Wind has its most important effects on crop production indirectly through the transport of moisture and heat. Vegetative growth at `Zero' wind, as experienced in glass houses or under low glass cover is luxurient. But, there is typically a reduction in vegetative growth as the wind increases to small values, viz., 1 or 2 metres per second.

Beneficial effects of winds

- Moderate turbulence promotes the consumption of carbon-dioxide by photosynthesis
- Prevent frost by disrupting a temperature inversion
- Wind dispersal of pollen and seeds is natural and necessary for certain agricultural crops and natural vegetation also.

Harmful effects of winds

- At sustained high speeds (12-15 metres per second) at plant height, plants assume a low, dwarf like form, whilest the intermittent high wind speeds experienced in gales and hurricanes results in gross physical damage to bushes and trees
- At higher wind speeds, the shape of the orchard tree alters giving rise to the characteristic wind shaping of trees in exposed positions
- Leaves become smaller and thicker
- Breakage occurs and bushes and trees subjected to natural (seasonal) pruning

- Direct mechanical effects are the breaking of plant structures, lodging of cereal crops or shattering of seed from panicles.

Management of High winds

- The effects of wind on evaporation can be avoided by using proper method of irrigation
- Crop management through crop rotations and inter cropping
- The damaging effect of wind can be reduced over a limited area by the use of shelter belts (rows of trees planted for wind protection) and wind breaks (any structure that reduce the wind speed)
- Crop varieties resistant to lodging
- Avoid fertilizer application to the crop when wind speed is greater than 20 knots.

3.5.3 Thunderstorms, Dust Storms, Hail Storms and Cloud Bursts

- These storms are known as local severe storms. It is estimated that at any given instant more than 2000 thunderstorms are taking place around the world.
- A thunderstorm is dreaded for the hazardous weather elements associated with it, such as lightning, hail, strong horizontal winds with shear and heavy rain
- These are meso scale phenomena with horizontal dimentions of 1-10 km and with a life span of a few hours
- A typical thunderstorm is made up of a single cumulonimbus cloud
- These are formed in a situation where a great deal of the energy for their genesis and development comes from the release of the latent heat of condensation in rising humid air
- These local storms cause severe damage to the standing crop by causing mechanical injury to the plants
- In dust storms, the dust is raised by the wind, covers small plants, which may cause stomata closure and suffocation
- Any thunderstorm that produces hails which reach ground is known as hail storm. Their diameter ranges from 5 mm to 15 cm and weight upto 0.5 kg. Unlike ice pellets hailstones are layered and irregular and clumped together
- Cloud burst is a rain gust or rain gush is a sudden heavy downpour over a small region. It is a remarkably localized mesoscale phenomena with rainfall of more than 100 mm per hour, area not

exceeding 20-30 square km with strong winds and lightning. Cloud burst represents cumulonimbus convection in conditions of marked moist thermodynamic instability and deep, rapid dynamic lifting by steep orography. Most of the damage to property, communication systems and human casualities result from the flash floods that accompany cloud bursts

- Hails cause direct damage to crops by lodging and shattering of seeds depending on their intensity.

Management of storms

- Prevention of hails by hail suppression techniques
- Afforestation to protect the soil from dust storm
- Use of organic fertilizers
- Establishment of tree belts and wind breaks
- Leaving the stubbles in the field
- Altering spring crops with winter crops
- Cover the orchards with hail nets
- Following forecasts of weather and protecting crops
- Spraying of salt on harvested paddy, to prevent the germination / sprouting of harvested produce.

3.5.4   Excessive or Defective Insolation

- Excessive solar radiation results in rise of soil and air temperatures. Defective insolation with consistent cloudy weather on one hand and consistantly bright and high intensity sunshine on the other hand causes enormous damage to crop plants.
- Cloudy weather retard growth, affect pollination and cause disease and pest incidence
- High solar radiation intensity cause pollen burst or flower drop.

Management

Since, these are very rare, the location specific solutions like:

- Proper site selection
- Allowing air drainage
- Adequate water supply
- Pruning of orchard trees
- Spray of chemicals and plant harmones

- Covering plants with "Hot caps" (covering plants with some standard and recommended material) may prove beneficial.

3.5.5   Tornado

- This is a violent, destructive storm of long vertical and small horizontal dimensions. A cumulonimbus cloud forms into a funnel shape with a vortex extending from the base of the storm to the surface. The whirl-wind encireles a small dimension of about 500 metres. These are capable of causing severe structural and other damages. The violent winds associated with this abnormality are strong upward air currents. The tornados occurring on water are known as "Water spouts".

Management

- Warning in advance
- Precautions to protect the agricultural produce like transportation to safety places
- Quick removal of debris immediately after damage.

3.6   Agrometeorological Services for Disaster Management

3.6.1   Preamble

Agriculture is defined as "The production and processing of plant and animal life for the use of human beings" and also "A system for harvesting or exploiting the solar radiation". Meteorology is "The science of atmosphere" and "The study of those aspects of meteorology which have direct relevance to agriculture" is defined as agricultural meteorology for which the abbreviated form is "Agrometeorology". The primary aim of agricultural meteorology is to extend and fully utilize the knowledge of atmospheric and other related processes to optimize sustainable agricultural production with maximum use of weather resources and with little or no damage to the environment. This entails improving the quantity and quality of agricultural crops and animal products and by-products. This science also provide the necessary information and attempts to reduce the negative impact of adverse weather (disasters, risks and uncertainties) through agrometeorological services. Agrometeorological services are "All agrometeorological and climatological information that can be directly applied to improve or protect agricultural production (yield quality, quantity and income obtained from yields) while protecting the agricultural production base from degradation. The current status of agricultural production and increasing concerns with related environmental issues call for improved agrometeorological services for enhancing and sustaining agricultural productivity and food security around the world. The disasters, risks and uncertainties in agriculture refers to the nature of weather and climatic hazards and damages and their potential impacts on loss of crops, animals, land and produce. Agrometeorological services play valuable part in making daily and seasonal farm management decisions and in management of disasters, risks and uncertainties pertaining to agriculture.

These services help the farmers in reducing the impact of disasters, risks and uncertainties in addition to efficient management of pests and diseases on their crops thereby help to increase their agricultural production. Agrometeorological services are useful in crop management systems that extension services provide to agricultural community and help the extension personnel in performing their functions more efficiently at the end user level because they work in centers where general agrometeorological information is potentially useful. The research scientists working in collaboration with different departments as multi-disciplinary team work for the right mix of knowledge to evolve agrometeorological products and services for farmers in different farming systems. These include forecasts of weather and climate, monitoring and early warning, products for drought, floods or other calamities and general agrometeorological advisories. These products and services would increase the preparedness of the farmers well in advance to cope with disasters, risks and uncertainties. In education the successes, failures and experiences of researchers and extension specialists are taught in the curricula of agrometeorology departments in the universities, training centers and other capacity building institutions. Such exercise will enlighten the classical training and strengthen the usefulness of the services for the farmers and other user communities. These agrometeorological services help the government decision makers to involve the private sector on priority in issues like crop insurances, providing infrastructure facilities, enhance cooperation between the institutions providing information and relevant advisories and those responsible for their transfer to the farming community to solve the risks and uncertainties associated with production agriculture and animal husbandry.

3.6.2 The role of Indigenous Technical Weather Knowledge (ITWK) in Agrometeorological Services

The indigenous knowledge has immense potential to manage the disasters, risks and uncertainties like climate change and variability, floods, cyclones, droughts and pests and diseases on crops and animals through agrometeorological services.

The use of *calotropis* to control thrips and mealy bugs, the use of cow dung cake gas as burrow fumigant and use of bow traps to control field rats, use of leaf powder of Margosa (*Azadiracha indica*), Nicotine (*nicotiana tobaccum*) and extract of custard apple (*anona squamosa*) for pest control on crops and animals and storing the grains in wooden bins. In addition, different species and varieties of crop seeds are mixed and sown to delay the onset and spread of pests and diseases, which also modify the micro climate for better harvest of all crops. The Himalayan farmers maintain their own rangelands plus a share in village managed community rangelands. They rare the animals and use cow and buffalo milk and milk products during failure of crops in drought situations. The farmers of China in mountainous regions follow irrigated (level) dry (bench) and complex (bench + level) terracing to manage water scarcity for crops. Similarly, farmers in Pakistan follow the same practices to derive the same benefits in addition to sustaining the productivity of mountain soil against landslides. The "Three north" system of shelter belts was evolved a few thousands of years ago and being practiced till date in China. This system protects the existing forests and rangelands against flood damages.

An indigenous system called *waru-warus* followed in Peru for over 3000 years. This system consists of a platform of soil surrounded by ditches filled with water. During droughts the moisture from canals slowly ascends through the roots in capillary action and during floods the furrows drain away excess run off. This system produces bumper crops even in the events of both droughts and floods. In Mexico a low cost and self sufficient farming system called *chinampa* practiced for centuries. In this system small plots of land are prepared which are separated by water channels. The water in the channels help to grow fish and also useful for irrigation (when in excess) and when scanty (drought) plants grown in the sides of *chinampas* gives income to the farmers. The "Sami" are the indigenous people in the northern Scandinavia. They live in Sweeden, Norway, Finland and Kola peninsula of Russia. They are a 70,000 strong population of which 16% of 17,000 Sweedish Sami are still reindeer herders who live in close contact with nature. They are not nomadic because the reindeer herding is their culture. However, they follow the path of reindeer, an agrometeorological service, between summer grazing lands in mountain regions and winter grazing lands in forests.

3.6.3 The Role of Contemporary Technological advances in Agrometeorological Services

Weather and climate data systems for agricultural activities are necessary to expedite the generation of products, analyses and forecasts to combat and evolve the preparedness measures against the natural disasters, management of risks and uncertainties in agriculture and environmental protection. The following products, tools and services of contemporary science and information technology have been providing newer dimensions to effectively monitor and manage the weather and climate related disasters, risks and uncertainties.

3.6.3.1 *Satellites and Remote Sensing*

One of the most important sources of agrometeorological data that compliments traditional methods of data collection is remote sensing. The satellite remote sensing is a new technique which provides spatial coverage of earth's surface and surrounding atmosphere. Use of remote sensing for rainfall estimation and monitoring the progress of cropping season are the needs of the hour across the globe for food security. Recent advances in satellite and computer technology had led to significant progress in remote sensing and proved its potential to meet the requirements of farmers at operational level.

3.6.3.2 *Geographical Information Systems (GIS) and Geographical Positioning Systems (GPS)*

The term GIS refers to a description of characteristics and tools used in the organization and management of geographical data. The term GIS is currently applied to computerized storage, processing and retrieval systems that have hardware and software specially designed to cope with geographically referenced spatial data and corresponding informative attribute. GIS enables management of large datasets such as traditional digital maps, databases and models. The quantitative data handling capability offered by GIS would assist the users to overlay numerous spatial data sets and statistically analyse the same. Through this procedure it is possible to develop quantitative relationships which are not achievable through the use of simple map drawing or graphics display programmes. GIS technology is a powerful agrometeorological tool for combining various map and satellite information sources in meteorological and climatological models that simulate interactions of complex natural systems. Therefore, it is possible to prepare decision support systems based on GIS to manage weather related disasters, risks and uncertainties. These tools would be handy for appropriate planning, coordination and monitoring of these events. The expected cyclone characteristics, rainfall pattern, water levels in rivers, environmental impact assessment and vulnerability, can help in modelling disaster consequences accurately and to evolve effective decision making on logistic and infrastructure requirement and their development in an area. The ultimate use of GIS lies in its modeling capability, using real world data to represent natural behaviour and to simulate the effect of specific processes. With this scientific background the frost risk map and the desertification climatic index were developed using spatial modelling with GIS, at the Portuguese Institute of Meteorology. On the same lines several information products (soils, crops and meteorology) were integrated in ISOP project in France which is in operational use and it assesses real-time forage production over France. The prediction and management maps of chilling injury on banana and litchi trees were developed using GIS in Gunangdong province, China which are highly successful and are being used by the farmers. A new major programme LANDFIRE uses GIS technology to map all wildfire fuels across USA at 30 m spatial resolution. This is intended to be the safety net for land management agencies. A few studies have focused on the collection of historical data and habitat conditions with the dynamics of locust development stages, and synthesis of data using GIS and evolving decision support systems. This system integrates remotely sensed soil texture, soil moisture and vegetation density with the daily weather data to forecast the suitable breeding sites and time of onset of locust upsurge in and around study area. Reliable drought interpretation requires a GIS based approach, since the topography, soil type, spatial rainfall variability, crop type and variety, irrigation support and management practices are relevant parameters.

The conventional methods of surveying and navigation require tedious field and astronomical observations for deriving positional and directional information. The GPS service consists of three components, viz., space, control and user. Rapid advancement in higher frequency signal transmission and precise clock signals along with advanced satellite technology have led to the development of GPS. The outcome of a typical GPS survey includes geocentric position accurate to 10 m and relative positions between receiver locations to centimeter level or better. The capabilities of surveying, mapping, locating geophysical positioning of GIS immensely help in combating the natural disasters in agriculture and effective control of environmental pollution.

3.6.3.3 *Information Technologies and Communications*

It is important that the user gets the information on agrometeorological services which are easily understandable and in time through a quick communication system. Effective communication and information about disasters, risks and uncertainties is a major challenge in the developing and under developed countries. Reliable communication networks connect the scientific and technological advances of the developed countries with these nations. The internet, digital satellite technology, wind-up machines and computers are new possibilities for rural areas in the under developed and developing world. The mobile phones, facsimile, e-mail and wireless technologies which are available in the developed countries offer greatest potential and must be recommended for the developing countries and under developed countries. Internet can accomplish accurate, timely, useful and cost effective   information to the rural areas. The RANET (Radio and Internet) system is an innovative system which brings new communications and technologies together and deliver operational agrometeorological services on disasters, risks and uncertainties of weather and climate over a distributed network in Africa. This is managed by the local communities and is rated as the most successful communication network for efficient communication agromet services. In India weather forecast bulletin for the subsequent three days is disseminated bi- weekly to AAS units on every Tuesday and Friday over telephone, telefax or satellite based very small aperture terminal (VSAT) communication system. The VSAT system has the capabilities for reliable interactive data communication and picture transmissions. Also dissemination of short and medium range forecasts to the farmers is being done through radio and television and the long range forecast regarding onset of monsoon, seasonal rain are disseminated through newspapers. The newly developed agrometeorological techniques are being communicated through the extension workers who reach the end users directly. The introduction of the satellite based cyclone warning dissemination system in 80s in Andhra Pradesh state of India was the single most important step to improve the speed and credibility of transmission of warnings for operational use by the farmers. The farming communities in a developed country like Germany are interested in receiving the longest term and most exact weather forecasts possible, as well as information about the expected conditions at the production sites. Therefore, telefax, T-online and Agromet- online services are in operation. The telefax allows transfer of much more information than publications and telephone, especially on agrometeorological forecasts. The T-online allows interactivity and Agromet-online includes the activity of farmer for he can take a relevant simulation into account when making his decisions.

3.6.4 Strategies to Improve the Agrometeorological Services to Cope with Risks and Uncertainties

The agrometeorological information plays a valuable part not only in making daily and seasonal farm management decisions but also in the management of disasters, risks and uncertainties. Globally, the current status of agriculture production, environmental degradation and the related sectors are influenced by the events of increasing disasters, risks and uncertainties. This situation calls for understanding and exploiting the agrometeorological information for the benefit of these sectors to take counter measures and to reduce the negative impacts of these events. It may not be possible to prevent the occurrence of these events, but the resultant negative and disastrous effects can be reduced considerably through the agrometeorological services. Therefore, the role of agrometeorological services and advance planning for the management of disasters, risks and uncertainties is crucial for these sectors because of the significance of their impact and influence on the overall well being of humanity and livestock. The agrometeorological services play a key role in strategic and tactical planning and efficient monitoring of crops and there is a growing recognition of the importance of operational agrometeorological services in all the sectors. In the summary and recommendations of the international workshop of WMO/CAgM, Accra (Ghana) in February 1999 on "Agrometeorology in the 21st Century – Needs and Perspectives" and during the subsequent 12th session of CAgM , the over all priority with ten agrometeorological services were identified. In addition, four types of support systems viz., (a) Data, (b) Research, (c) Education/training/extension and (d) Policies were distinguished. These support systems have to be used operationally for agrometeorological applications in agriculture, forestry, rangeland, environmental protection and other related sectors to reduce the impacts of disasters, risks and uncertainties in these sectors. Four years later, the

13th session of the CAgM of the WMO, held in October, 2002 in Slovania, considered the need to improve the agrometeorological services in addition to the support systems to increase agricultural production, livestock and to conserve the environment. These tasks could be achieved as detailed below:

3.6.5 Improving the Agrometeorological Services

To cope with the disasters, risks and uncertainties pertaining to agriculture, forestry, rangeland, environment and livestock, the list of actual agrometeorological services from the above workshops and the CAgM literature and the strategies to improve the same are:

3.6.5.1 *Agrometerological characterization, using different Methodologies*

The purpose of agroclimatic characterization is to identify those aspects of climate which distinguish a region from the nearby regions and to draw inferences on the influence of climatic factors on crop production. The hypothesis is that under given climatic conditions there are similarities in crop growth and development thereby the yield in that homogenous region. Four decades ago itself it was observed that the failures or disappointing results of agricultural development projects in various parts of the world including projects to produce pineapples in Philippines, sugar in Puerto Rico, peanuts in East Africa and rubber in Amazon basin may have been largely due to failures in proper agroclimatic classification. It was also stressed that adequate assessment of agroclimatic resources is an essential prerequisite for proper planning of agricultural development. The homogenous crop zone boundaries with relevant crops and cropping patterns could be delineated with different methodologies through appropriate and scientific agrometeorological characterization. The resultant agroecological zoning offers the potential for developing strategies for efficient and sustainable natural resource management, including sustainable management of agriculture, forestry, rangeland, environment and livestock. In addition, agricultural risk zoning is must and form as an essential component of natural disaster mitigation and preparedness strategies. Given the complex nature of databases, the GIS and remote sensing should be employed in future in such studies to facilitate strategic and tactical applications at the farm and policy levels.

3.6.5.2 *Advice on design rules on above and below ground microclimate management or manipulation, with respect to any*

appreciable microclimatic improvement; shading, wind protection, mulching, other surface modification, drying, storage and frost protection

The climate of a region determines the extent of adaptability of a crop (or animal) species and weather influences its day to day growth. In turn the crops not only modify their own microclimate and weather within their canopies but also the soil underneath them due to emission of long wave radiant energy. Any modification in soil, agricultural practices or ground cover vegetation may have consequences for the carbon cycle *via* their impact on the dynamics of soil organic matter. Conversely, changes in composition of atmosphere may bring about changes in certain soil characteristics. (Example: Parkland agroforestry and other stabilizing intensive management of scattered or clumped or allayed trees for microclimate management and manipulation to cope with temperature changes in northern China plain). It is suggested that there should be more research at micro level into the physical behaviour of crop growth like profiles of solar radiation, temperature, wind speed, vapour pressure, carbon-dioxide demand and moisture regimes to develop better agricultural mitigation strategies at micro level against risks and uncertainties. Micrometeorological knowledge about energy exchange and transports at the surface has useful applications in agriculture. The changes on a small scale are relatively easy to initiate and control. Examples: Studies also on increasing surface absorptive power, exposure through site selection, artificial or natural shading for reduction of day length.

3.6.5.3 *Advice, based on the outcome of response farming exercises, from sowing window to harvesting time, using recent climatic variability data and statistics or simple on-line agrometeorological information*

Response farming is defined as "A method of identifying and quantifying the seasonal rainfall variability and predictability to address the problem of the farmers at field level". However, it was suggested that response farming should not only be considered with respect to fitting cropping seasons to variable rainfall patterns but also for temperature patterns. A case study from Vietnam shows that either a planting date or a combination of planting date and variety could be varied to make sure that the rice flowers optimally with the detailed knowledge of temperature. Response farming has become a promising technology in the past two decades to alter cropping systems/patterns, in relation to fluctuations in seasonal weather. Therefore, it is suggested that in response farming where ever the indigenous technical knowledge is available that shall form a part in finding solutions to farming problems by improved use of available forecasting in the cropping season(s).

3.6.5.4   *Establishing measures to reduce the impacts and to mitigate the consequences of weather and climate related natural disasters for agricultural production*

The plan of implementation of the World Summit on Sustainable Development (WSSD) held in Johannesberg in 2002 highlighted the need to mitigate the effects of disasters, risks and uncertainties. They suggested the measures such as improved use of climate and weather information and forecasts, early warning systems, land and natural resource management, agricultural practices and ecosystem conservation. The WSSD noted that these measures would reverse the current trends and minimize the degradation of land and water resources which are the basic needs for agricultural production. Therefore, there is a need to promote the access and transfer of technology related to early warning systems and mitigation programmes to developing countries affected by risks and uncertainties. A comprehensive documentation of risks and uncertainties related to agriculture and allied fields at national, regional and international levels is very important. This process helps to develop mechanisms for more efficient assessment of the impacts of the risks and uncertainties in all fields in general and agriculture in particular. In response to the importance of natural disaster reduction, there is a pilot project for a Joint Programme to Contribute to Natural Disaster Reduction in Coastal Lowlands.

3.6.5.5   *Monitoring and early warning exercises directly connected to already established measures*

The role of early warning and advance planning for natural disaster management and the mitigation of extreme weather/climate events is crucial for agriculture, forestry, rangelands, environment and livestock. The application of weather and climate information to improve the effectiveness and efficiency of emergency preparedness and response activities is essential. There is also a need to monitor critical thresholds that should trigger early warnings. So, it is essential to survey the status of trends in land degradation and to report on appropriate criteria to conserve and manage material and environmental resources for the benefit of these sectors. Rapid advances in information technology need to be rapidly transferred to operational applications to more effectively disseminate agrometeorological information to the user community. There needs to be full involvement of all users of the information as well as the providers of the information to ensure that the right information is delivered to the right user at the right time for early warnings. There is an urgent need to identify the information gaps and establish guidelines and procedures to improve the flow of timely and accurate information to farmers, including both monitoring and early warning systems. Current natural disaster management is largely crisis driven. There is also an urgent need for a more risk-based management approach to natural disaster planning in agriculture, rangelands, forestry, environment and live stock. The concept of the drought monitor map product be promoted as a tool for all drought prone countries to better understand drought severity using multiple indicators. The feasibility of organizing joint training workshops on national and regional drought monitor products under the auspices of WMO and the NDMC should be examined. Indices used in China in their agrometeorological bulletin could be effective training tools.

3.6.5.6 *Climate predictions and forecasts and meteorological forecasts for agriculture and related activities, on a variety of time scales, from years to seasons and weeks, and from a variety of sources*

Weather forecasts play an important role in agriculture because they help in agriculture planning in advance. Therefore, the study of climatic fluctuations in the rainfall and their impact on agriculture has become an important area of climatology in recent decades. When the average distribution of weather and climate phenomena are studied along with the average variations in frequency and extent, then it gives a better insight into agronomic importance. So, one of the persistent demands of agriculturists is for reliable forecasts of seasonal climate information to make appropriate decisions as to which crops and cropping patterns to choose well ahead of the growing season in order to avoid undue risks and uncertainties. Hence, there is also an urgent need to assess the forecasting skills for natural disasters to determine those where greater research is needed. Lack of good forecast skill in drought, for example, is a constraint to improved adaptation, management and mitigation. For the identification of climatic fluctuations in rainfall data one needs long and continuous records. Such data series are available only at few stations in developing countries. Different techniques like relative rainfall probabilities, moving average, iterative auto-regression may be adopted for better defining the climatic fluctuations, cause and effect of such fluctuations and expected fluctuations to food producing ecosystems.

3.6.5.7 *Development and validation of adaptation strategies to increasing climate variability and climate change and other changing*

conditions in the physical, social and economic environments of farmers

Scientific assessments have shown that over the past several decades, human activities, especially burning of fossil fuels for energy production and transportation are changing the natural composition of the atmosphere. Providing adequate standard of living (adequate food, water, energy and healthy environment) for the current and future generations is a major challenge. So, there is a need in many agricultural areas around the world to enhance the understanding of climate variability in order to assess the impact of causal factors (natural and human). A better understanding of the climate of the major ecosystems of the world where agriculture and related sectors are at risk could help develop effective *in situ* coping strategies. There is a need for thorough understanding of the effects of changes in regional climate on crop production, forestry, rangelands, environment and livestock. Given the growing incidence of dust and sand storms around the world, it is essential to include measurement of aerolian sedimentation loads in the standard agrometeorological stations of NMHSs. It is also essential to include a routine and comprehensive analysis of wind speed and direction data and disseminate this information to the users. These data should be applied to analyze the impact of sand storms on agriculture. The issue of distinguishing long-term climate variability (eg., IPO) and long term climate change is important, as is the need to consider the impacts on agriculture, water resource management, and disasters such as bushfires. This is important because there are implications for long term sustainability of certain types of activities, especially agriculture. There is modeling work at more overseas institutions (e.g., the Hadley Centre) that would be of relevance here. These issues need to be drawn to the attention of national policymakers.

3.6.5.8 *Specific weather forecasts for agriculture, including warnings of suitable conditions for pests and diseases and/or advice on mitigation measures, such as fire weather monitoring*

In natural ecosystems and also in cultivated or forest ecosystems climate change is capable of disturbing the balance between the species, whether they are plant or animal in terms of individual and population. The effect of climate changes on development of pests and diseases could manifest a direct effect on biological cycle of parasites and host parasite interaction. Weather influences the degree to which plants and animals (i.e., the 'victim' and/or 'host) are attacked by pests and diseases or harbour them. It also affects the biology of insects and disease organisms, and determines the nature, numbers and activity of pests (and of predators on pests) and extent and influence of diseases. In crop and livestock protection, the spread and aerial transport of pests and diseases and the effectiveness of applied control or eradication methods depend upon atmospheric agencies. Agrometeorologists need accurate and reliable climate forecasts to assist the agricultural community in planning and operations. There is a need to appraise appropriate authorities of current capabilities in the analysis of climate change/variability and long range prediction studies as they specifically relate to agriculture, rangeland, forestry, environment and livestock at the national and regional levels for appropriate applications. Given the need for guidance in natural disaster management, case studies of China and other countries in agriculture, rangelands, forestry, and fisheries and the application of storm surge forecasting at the local level must be documented.

3.6.5.9 *Advice on measures to reduce the contributions of agricultural production to global warming and on keeping an optimum level of non degraded land dedicated to agricultural production*

Agriculture in 21st century will have to make its contribution to the reduction of GHG emissions (particularly CO_2, CH_4 and N_2O) to satisfy the vital needs of populations in food, energy, fiber and other products. The adoption of agriculture to global warming trigger new requirements from major contemporary research efforts. This research shall aim both to increase the forecasting capacity and to anticipate the design of new cropping and forestry systems. Global warming in all sectors of agricultural production necessitates the adoption measures that must be economically feasible. Such measures could improve resilience of agricultural production systems to global warming, but, do not necessarily reduce emissions from the agricultural sector. To serve the agricultural sector, there is a need to thoroughly review the interactions between greenhouse gas emissions and agricultural activities. Also, there exist needs to document both positive and negative influences of agriculture on weather and climate systems and develop guidelines for increasing awareness within the farming communities of the related; adaptation/mitigation strategies to address global warming and poverty issues. More attention should be given to the impacts of potentially increasing frequency and severity of extreme events associated with global warming and appropriate mitigation strategies.

3.6.5.10 *Proposing means of direct agrometeorological assistance in the management of natural resources for the development of sustainable farming systems via technological advances with strong agrometeorological components*

While scientific and technological advances have resulted in higher quality information and increased capabilities in providing agrometeorological information, major difficulties remain. Technological advances in remote sensing (hand held and space based) such as soil moisture detection and evapotranspiration estimations and GIS constitute new sources of data for many agrometeorological applications and should help to reduce some of these problems in future. The RS and GIS data sets not only complement ground observations but also offer new types of data and also provide greater global coverage and improve aerial averaging. Regional and global cooperation can hold the countries that lack the financial and technical resources to acquire such data. As indicated in certain above sub heads, recent technological advances in GIS offer significant improvements in spatial analyses of meteorological and agricultural data bases. Sustainable research and development has occurred in the application of crop models ranging from the field level to country level and even larger scale modeling. Various modeling techniques range from statistics based regression analysis to more complex process oriented approaches. Models are also used in global changes impact studies. The problem is how to develop an integrated information management system with computer technology and standardized analytical techniques that can be applied operationally for validation for selected models in agriculture, rangelands, forestry and environment at the ecoregional level. Therefore, it is recommended that an integration of GIS, remote sensing, simulation models, and other computational techniques be used to develop more effective early warning alerts of natural disasters, risks and uncertainties. Also, there is a need and opportunity for the agrometeorologists to supply design requirements for new satellite sensors. This applies in particular to droughts, rangeland management and to combat forest fires from disaster mitigation point of view.

To cope with the risks and uncertainties the above Agrometeorological services shall be delivered not only at operational level but also at the strategic and tactical levels where planning of agricultural operations, both short and long term are included.

3.6.6   Improving the Support Systems of Agrometeorological Services

To cope with the disasters, risks and uncertainties pertaining to agriculture, forestry, rangeland and environment the agrometeorological support systems (data, research, education/training/ extension and policies) from different workshops and the CAgM literature and the strategies to improve the same are detailed below:

3.6.6.1   *Data*

At present collection, management, and analysis of atmospheric and surface data are being done with automated instruments. The new techniques like remote sensing and GIS cover the data from near ground to outer space. However, these data support systems have grown only in the developed countries. To overcome the difficulties faced by the developing and under developed countries the cooperation among international, regional, national and where possible local specialized bureaus and organizations is recommended. Assessment of the impact of natural disasters on agriculture, rangelands, forestry, and environment requires the design of a comprehensive data base in accordance with the user needs. There is a need for an integrated data management system, from adequate collection to quality control, analysis, presentation and also metadata not just meteorological data. Presentation should make use of best available technology, e.g., GIS and Internet. Effective management of, and preparedness for natural disasters, risks and uncertainties require free and unlimited access to relevant databases that will allow monitoring, assessment and prediction. It is recommended that all agencies responsible for these databases develop good collaborating links for the exchange of information included in these databases.

3.6.6.2   *Research*

This agrometeorological support system needs regional, national and local prioritization. The priority items identified at Accra (1999) are:

- Agrometeorological aspects of the efficient use and management of resources in full production environment
- Reduction of impacts on resource base, yields and income from natural disasters, risks and uncertainties
- Validations and applications of databases and models for well specified systems and users
- Ways to ensure that research results are adopted in farming.

These aspects are already the priority areas in OPAGs of CAgM and needs further vigorous persuasion. There is an urgent need to assess the forecasting skills for natural disasters to determine those where greater research is needed.

3.6.6.3   *Education/Training/Extension*

With increasing incidence of natural disasters, risks and uncertainties around the world, a comprehensive assessment of their impacts on agriculture, forestry, rangelands, environment and livestock and strategies for mitigation of natural disasters is critical for sustainable development, especially in the developing countries. Education and Training is an important component in these sectors. It is recommended that strategies for education and training address the needs at national, regional, and international levels in order to exploit the synergies and share experiences. Community involvement and education is essential in preparedness and mitigation.

3.6.6.4   *Policies*

An appropriate policy environment based on social concerns and environmental considerations can help develop the right mix of strategies for preparedness and problem solving practices against natural disasters, risks and uncertainties. It is recommended that countries develop policies aimed at effective natural disaster management. Such policies should emphasize incentives over insurance, insurance over relief, and relief over regulation. The growing frequency of natural disasters require effective use of the media to better inform and educate the general public and policymakers about the potential impacts of natural disaster and the need to adopt the preparedness strategies. Given the regional and global nature of natural disasters, it is essential to promote and foster collaboration between agencies and between international and regional programs and build partnerships.

3.7 Conclusions

Agrometeorological services are the end products of agrometeorological research. Improved management of climate and weather related events such as disasters, risks and uncertainties is central to the profitability of rural industries and the ecological sustainability of its resource base. Agriculture, forestry, rangelands, environment and livestock are the most important sectors heavily impacted by these events. The current status of agricultural production and increasing concerns with related environmental issues calls for improved agmeteorological services for enhancing and sustaining agricultural productivity around the world. The requirements for the agrometeorological services were described in the light of emerging issues related to environment, climate change, biodiversity, drought and desertification, food security and sustainable development. The Agenda 21, International Conventions including the United Nations Framework Convention on Climate Change (UNFCCC), the Convention on Biological Diversity (CBD) and the United Nations Convention to Combat desertification (UNCCD), the World Food Summit Plan of Action and the World Summit on Sustainable Development (WFSPAWSSD) include the elements that have implications for strengthening operational agrometeorological services. The weather related disasters around the world have been increasing in the recent past and operational agrometeorological services could help the farming community with better preparedness and mitigation strategies. Towards this end some strategies to cope with these vents were discussed *in situ* in the text above and excerpt of the same is as follows:

- With the increasing incidence of events such as natural disasters, risks and uncertainties around the world, a comprehensive assessment of their impacts on agriculture, forestry, rangelands, environment and livestock and strategies for mitigation of the same is critical for sustainable development, especially in the developing countries

- Agrometeorological services and information must increasingly be made available to assist farmers in characterization of agroclimate, microclimate management and manipulation, advisories on response farming, monitoring of and early warning on natural disasters and establishing measures to reduce their consequences, climate prediction and forecasting along with forecasts for pests and diseases and management of natural resources

- There should be more research into the physical behaviour of crop growth and moisture regimes to develop better agricultural mitigation strategies and standardisation of services and products. The research shall also aim at improved agrometeorological services, not only for enhancing agricultural productivity, but also for protecting the environment and biodiversity, coping with climate change and drought and desertification for ensuring sustainable development

- Training programmes and education at regional levels which include aspects related to climate modeling, integration of satellite imagery oriented to agricultural GIS analytical tools shall form a regular feature. Regional exchange programmes that will consider the transfer of methodologies and knowledge of professionals of different services, by means of seminars, workshops and/or hands on training must be given top priority

- The regional and global nature of the natural disasters, risks and uncertainties and the complexity of issues involved demands to promote and foster collaboration between agencies and between international and regional programs and build partnerships. This process helps in increasing the network of agromet stations, maintain the existing ones and develop competitive agromet products

- New Initiatives such as the WAMIS could help strengthen operational agrometeorological services through the provision of agrometeorological products on a near real time basis on the internet and through training modules to enhance the quality of agrometeorological products. Priority shall also be given and enough funds allocation be made for dissemination of meteorological tools applied to agriculture oriented towards small and medium farmers.

4

Weather – Erosion – Remote Sensing – Crop Growth Models

We are all born with a divine fire in us. Our effort should be to give wings to this fire.

- APJ Abdul Kalam

4.1 Soil Erosion

Soil is one of the most important, critical and precious natural resources of planet earth. Nearly 99% of human food comes from land. So, it is a source of human sustenance and food security. Economic stability and wise use of soil for agriculture are inseparable across the globe. Therefore, Judicious management of soil is necessary for sustaining agricultural productivity and environmental security. The soil degradation is due to erosion (water and wind), water logging (flooding), salinity (alkalinity), soil acidity, complex problems etc., among which, most soils in the world suffer severe degradation due to erosion. Soil erosion is as old as agriculture. At normal geological pace, nature requires 1000 years to build up 2.5 cm of top soil, but wrong farming methods may take only a few years to erode it from the lands of average slope. A fraction of a millimeter (about 1 tone soil 1 annum) is added on to 1 ha though the loss of 100 t/ha/year is not uncommon from undulating terrain.

Definitions

"Soil erosion" is a process of detachment and transportation of soil particles from the top soil by wind and or water. There are two types of erosions viz., Geological and Accelerated. The geological erosion is natural or normal and is said to be in equilibrium with the soil forming process. It takes place under natural vegetative cover completely undisturbed by biotic factors and the present topographic features like stream channels and valleys are developed through this erosion. On the other hand the accelerated erosion is due to disturbance in natural equilibrium by the activities of man and animals through land mismanagement, destruction of forests and over gazing. The soil loss through erosion is more than the soil formation due to soil forming processes. Wind and water are the main agents responsible for soil erosion. The loss of soil from land surface by water, including run off from melted snow and ice is usually referred to as "Water erosion". The process of lifting and blown off of the single grained soil particles on the land is known as "Wind erosion". Soil erosion caused by water can be distinguished mainly into (a) Sheet erosion (b) Rill erosion and (c) Gully erosion. The erosion in which soil matrix is lost but remains undetached for long period and removes a thin layer of soil from large areas uniformly during every rain event producing a run off is called as "Sheet erosion". When sheet erosion is allowed to continue unchecked, the silt ladden run off form a well defined minute finger shaped grooves over the entire field, and the process of formation of grooves is known as "Rill erosion". When rill erosion is neglected, the tiny grooves develop into wider and deeper channels known as "Gully erosion". The wind erosion mechanism and process involve (a) Saltation (b) Suspension and (c) Surface creep. The major portion of the soil carried by the wind moves in series of short bounces called "Saltations" carrying fine particles of 0.1 to 0.5 mm in diameter. The movement of the dust particles, smaller than 0.1 mm diameter by floating in the air is known as "Suspension". The rolling and sliding of soil particles along the ground surface due to impact of particles descending and hitting during saltation is called "Surface creep". Coarse particals larger than 0.5 -2.0 mm diameter are moved by surface creep.

4.1.2   Role of Weather Elements in Soil Erosion

Rainfall (Water)

Erosion by rain is a natural process on slopping land. It is the detachment and down slope displacement of soil particles by rain drop impact (splash) and running water (overland flow rain) enhances the translocation of soil through the process of splashing. Individual raindrops detach soil aggregates and redeposit particles. The dispersed particles may then plug soil pores, reducing water intake. Once the soil dries, these particles develop into a crest at the soil surface and runoff is further increased. When the natural vegetation is removed as in arable farming, the soil is exposed to rainfall and the rate of erosion may increase 100-1000 fold, compared to natural vegetation.

The effectiveness of conservation tillage for erosion control is based on keeping the soil surface covered with growing crop and or crop remains at all times in order to prevent rain drop impact forming a soil surface crust that would hinder infiltration of rain water, thereby giving rise to overland flow.

Wind

Wind erosion occurs where soil is exposed to the dislodging force of wind. The intensity of wind erosion varies with surface roughness, slope and type of cover on the soil surface and wind velocity, duration and angle of incidence. Wind erosion physically removes the lighter, less dense soil constituents such as organic matter, clays and silts. Thus, it removes the most fragile part of the soil and lowers soil productivity. The susceptibility to wind erosion is related to water content, stability of dry soil aggregates, stability of soil crust, surface roughness, vegetative and or mulch cover of the soil. The long unsheltered and smooth soil surfaces are prone to wind erosion during critical periods of the year.

4.1.3   Erosion Control and Adaptive Measures

4.1.3.1   *Asia and Oceania*

According to GLASOD assessment, a total of 83 m ha is assessed as affected by water erosion in the region. This is made up of 33m ha with light erosion, 36 m ha moderate and 14 m ha strong erosion. The dry zone is most affected with 39% of the area under crops and pasture compared with 18% for the humid zone. In terms of absolute area, the countries most seriously affected are India and Iran. However, relative to crops and pasture the countries effected are Iran, Srilanka and Nepal. The areas where erosion has reached the severe degree, leading to abandonment of land include parts of the hill areas of Sri Lanka and the Pothwar plateau of the Punjab region of Pakisthan.

In the GLASOD estimate, a total of 59 m ha is assessed as affected by wind erosion in the region lying entirely in dry zone. Agricultural land accounting 60% in Iran and 42% in Pakistan are effected by this erosion. Most of the wind erosion occurs in the dry belt stretching from Central Iran to the Thar Desert of Pakistan and India. Relatively low proposition of Afghanistan is effected by wind erosion.

The Central Soil and Water Conservation Research and Training Institute (CSWCRTI) Dehradun, India conducts research on rainfed biomass productivity and resource conservation. A brief summary of the research results and accomplishments are given below:

- Non monetary inputs like contour cultivation, early sowing, cover management of row crops through legume inter cropping and narrow spacing (90 × 20 in maize) for reducing runoff and soil loss and increasing production on moderately slopping lands were devised

- Organic mulching @ 4 t/ha effectively reduced erosion losses from 37 to 6 t/ha on 8% slope soils of Doon Valley and also increased yield of the subsequent crop from 1.9 to 2.4 t/ha

- Normal tillage with live mulching (green manure crop grown in between maize rows) reduced run off by 55% and soil loss by 60% at 4% slope

- Variably graded bund for annual rainfall greater than 600 mm in permeable soils and less than 600 mm in impermeable soils disposed excess water safely

- Bench terracing in 16 to 35% slopes of red and lateritic soils reduced runoff from 15 to 3% and soil loss from 45 to 0.5 t/ha

- Run off farming with water harvesting, storage, recycling, tapping of perennial flows and augmentation of well water for supplemental irrigation through conservation practices has been very well received by farmers

- Deep gullies greater than 9 meters were stabilized by constructing gully plugs, small earthern check dams, live hedges and staggered contour tranches.

As detailed above many measures, techniques, tools and products are developed by CSWCRT, Dehradun, India to prevent and reduce soil erosion for over 60 years. These research findings need to be adapted to local conditions of climate, soil, slope, field length and type of crop grown on the eroded fields.

4.1.3.2 *Africa*

Sahel is the region of West Africa which separates Sahara desert from the Savannah. Land degradation is threatening the economic stability of the region. The Sahel has degraded because large human and livestock populations and increased cultivation have reduced native vegetation in the area. As a result, excessive run off has increased erosion and decreased soil fertility. Sediment carried by the runoff is deposited into Niger river, where it creates alluvial fans (unproductive land forms made of transported soil and rock from the Sahel). Scientists devised a way to prevent further erosion and sediment deposits through reforestation using micro catchments. The micro catchments are crescent – shaped trenches, about 4 feet in length, built on plateaus in the path of erosion. The trenches catch and hold moving water and sediment, preventing the sediment from polluting the river. Trees and vegetation when planted in the trenches provide extra ground cover to further reduce erosion.

Soil erosion is one of the chronic environmental problems being faced by Zambia's Lusitu area in Gwembe Valley of Southern Province. Just few hours of torrential down pours can wash away tons of top soil from each hectare of land. More than 53,000 Tonga speaking people in the area are facing a bleak future. For the past 40 years, the valley's forests have been disappearing 30 times faster than they are being planted while hill sides are continually being stripped off their protective covering of vegetation. This problem was solved by planting vetiver (Khus Khus) a tropical grass, along the contours of sloping lands. It quickly forms narrow but dense hedges. Its stiff foliage then blocks the passage of soil and debris. It also slows any run off and gives the rainfall a better chance of soaking into the soil instead of rushing off the slope.

However, the vetiver is not new to Africa. In Kenya it had been used as ornamental plant. In Malawi, vetiver is said to have been used for over 50 years to stabilize sugarcane. The Mauritania settlers in Chirezi area, a tropical area, have reportedly used it for years to reinforce the bank of irrigation channels. It has also been reported that tobacco farmers have found hedges of vetiver around their fields keep out creeping grass weeds, such as "Kikuyu" and "Couch". In South Africa, vetiver is cultivated to a limited extent and is used as hedge plant, particularly in Kwazulu – Natal where it is used mainly by Mauritian sugarcane growers and is commonly referred to as "Mauritius grass". In Burkina Faso certain tribes rely on the plant for livelihood. For example, the Bozo tribes who dwell along the vetiver covered shores of Lake Niger weave their huts and baskets out of vetiver grass. The vetiver grass may eventually benefit watersheds, forests and farms from soil erosion by water and wind.

A field experiment was conducted in Southwest Niger on sandy, siliceous and isohyperthermic soil during early rainy seasons of 1994 and 95. Particle mass transport was quantified in two plots of 55 by 70 m. During the first storms of both seasons, the plots were without a mulch cover. Afterwards, one plot was covered with flat pearl millet (*Pennisetum glanceum (L.)*) stalks. The application rates were 1500 and 1000 kg per ha. The 1500 kg per ha mulch cover reduced sediment transport from 49.7 to 80.2 % during five storms with wind speeds varying from 8.3 to 10.6 m s^{-1} and is therefore recommended as the better application rate for wind erosion control in the Sahel.

4.1.3.3   *Europe*

In Europe Scientific interest in soil erosion arose in 1970s. This was due to changes in agriculture (larger fields, more low crops, heavier machines) leading to an aggravation of erosion. On a European level the recent study to encourage farmers to adopt soil conservation practices is the SoCo project. The soil that is washed away is deposited as mud on roads, in road side ditches, in culverts and sewers, in gardens, basements, and cellars of houses, in the streets of built up and residential areas and in rivers, dams and reservoirs. So, there may be serious damage to public and private property outside agriculture. Soil erosion is often also accompanied by flooding and water quality may be impaired by erosion derived agrochemicals. The mechanisms and controlling factors of soil erosion and of basic principles of soil conservation techniques that are effective and feasible in Europe are:

- Conservation tillage in its various forms (minimum tillage, zero tillage, no till, direct drill, reduced tillage, strip till, mulch till, non-inverting tillage) has become increasingly popular as a means to combat soil erosion

- According to EC AF, the agricultural conservation in Europe was very little developed in 1998 (estimated at < 1% - 2% of its agricultural land). France and Spain are the two countries in Europe where these techniques are practiced the most, with about 1 and 0.6 million hectares of annual crops under conservation techniques in 1998 and this is increasing

- The following data for 2003 is given by EC AF for the percentage of the agrarian under conservation agriculture. Belgium (10%) Ireland (4%) Slovagia (10%) Switzerland (40%); France (17%); Germany (20%); Portugal (1.3%); Denmark (8%); United Kingdom (30%); Spain (14%); Hungary (10%); and Italy (6%). A special form of conservation tillage i.e., non-inverting tillage with a type of plough that cuts a slice of soil loose from the sub soil without inverting it has become popular

- As of 3rd April, 2008 farmers in Dutch South Limbourg can apply for a subsidy of 50 Euro per ha per year for adopting non-inverting soil tillage. Starting 2013, conventional plashing was no longer allowed in the region South of Sittard

- It was reported that in a composite analysis of long term climatic controls on main erosivity in the Calore river basin (Southern Italy) it was indicated that soil erosion risk tends to increase as a consequence of an escalation of climate erosive hazard (increase in rain resulted in increased erosion)

- Wind erosion is not as significant or a wide spread problem in Europe as in dryer parts of the world. The hazard is greatest in the low lands of northwestern Europe with more than 3 million ha to high potential wind erosion risk. Most farmers only use measures to protect their high value crops. It was observed that tools like re-afforestation of arable land can help regional policy makers with the implementation of wind erosion control measures.

4.1.3.4   *North America*

Soil erosion by water and wind is a serious problem on approximately one million acres of land in California hills and valleys traditionally used for production of barley and wheat.

In California, many small grain producers use minimum tillage. They also reduce the tillage operations and adjust chisels, discs and cultivators to leave sufficient crop residue levels to qualify as conservation tillage. Most of the farmers adopt crop residue management methods including no till and this proved the most successful measure to control water and wind erosions. Contour farming and terraces, contour and strip cropping, contour terraces and grassed water ways and buffer strips proved significantly advantageous and successful agro metrological measures to protect land and mitigate land degradation by erosion in Northern America.

Wind erosion in the United States is most widespread on agricultural land states in the Great plains. This is also a problem on cultivated organic soils, sandy coastal areas, alluvial soils along river bottoms, and other areas in the United States. During the 1930's a prolonged dry spell culminated in dust storms and soil destruction of disastrous proportions. The "Black blizzards" of the resulting "Dust bowl" implicated great hardships on the people and the land. Nearly seventy years after the "Dust bowl" ended, wind erosion continues to threaten the sustainability of USA's natural resources. In the spring of 1996, wind erosion severely damaged agricultural land throughout the great plains. On cropland, about 70 m ha are eroded by wind and water at rates that exceed twice the tolerance level for sustainable production. On average, wind erosion is responsible for about 40 per cent of this loss and can increase markedly in drought years. According to the 1992 National Resources Inventory (NRI), the estimated annual soil loss from wind erosion on non federal rural land in the United States was 2.5 tons per acre per year. This number is a decrease from 3.3 ton per acre per year in 1982. Much of this reduction was a result of enrollment of land classified as highly erodible in the conservation reserve programme (CRP).

In a study on "Wind and water erosion under global change type conditions and alternate land management practices" it was observed that wind erosion exceed water erosion by a factor of 50. Grazing and fire amplify both water and wind erosion. The dust emissions likely dominate water driven sediment production under global change type conditions. It was emphasized that the need for more careful land management to avoid further amplification of dust emissions.

4.1.3.5 *Latin America and the Caribbean*

The Latin America and Caribbean region has the world's largest reserves of arable land with an estimated 576 m ha equal to almost 30 percent of the total territory. The region also contains 16 per cent of the world total of 1900 m ha of degraded land taking third place behind Asia and the Pacific and Africa. During 1972-99, the area of permanent arable land and cropland expanded in the South America by 30.2 m ha or 35.1 per cent in Meso America by 6.3 m ha or 21.3 per cent and in the Caribbean by 1.8 m ha or 32.0 per cent. Erosion is the main cause of land gradation in Latin America affecting 14.3 per cent of the territory in South America and 26 per in Central America. There are both a tendency to merge farms to make larger holdings and an increase in number of small holdings. Both the processes have adverse environmental effects. In large farms, the land suffers from erosion and compaction due to mechanization, as well as salination because of improper irrigation and chemical pollution. Small holdings increase deforestation and lead to erosion and loss of soil fertility because they are used intensively without allowing for adequate fallow periods.

4.1.3.6 *Technological Intaventions for Erosion Control*

Sources indicate that $1/3^{rd}$ of worlds arable land lost in erosion since 1950, mostly in Asia, Africa, S. America @ 13-18 t/acre/yr. This resulted in decreased organic matter and increased calcium carbonate.

Simultaneously, the microbial biomass reduced resulting in slower decomposition of plant residues. Therefore, conservation technology such as implementation of mechanical structures and/or zero-till or minimum tillage reduces the this huge amount of soil loss. However, a major factor is the period when the land is bare or devoid of vegetative cover and the number and intensity of storms at the onset of rainy season.

Water Erosion

Globally, there are about 56 million km^2 of land (43% ice free) vulnerable to water erosion in the slightly arid to humid areas of the world. Properly constructed and maintained waterways with good vegetative cover can be a practical way to prevent water erosion. Waterways must have a shallow, saucer shaped cross section and an erosion resistant vegetative cover to carry water safely. Further specific practices to avoid water erosion include:

- Growing forage crops in rotation or as permanent cover especially close to surface
- Growing winter cover crops
- Interseeding
- Protecting the surface with crop residue
- Shortening the length and steepness of slopes
- Increasing water infiltration rates
- Improving aggregate stability
- Increase the content of soil organic matter which helps improve soil structure
- Use contour furrows, terraces, plowed strips, and /or ridges to reduce or deflect runoff
- Land shaping/leveling/graded bunds and diversion channels.

Wind Erosion

Globally, there is about 33 million km^2 or 25% of land mass that is vulnerable to wind erosion. Wind erosion is a serious problem in many parts of the world. It is worse in arid and aqsemiarid regions. Areas most susceptible to wind erosion on agricultural land include much of North America and the Near East; parts of Asia (southern, central and eastern); The Siberian plains in Australia, northwest China, southern South America. Wind erosion must be fought in three areas viz., source areas where particles should be blocked, transport areas where winds need to be redirected to avoid human constructions and infrastructures from being buried accumulation areas where mobile sand must be stabilized. In addition, specific practices to avoid wind erosion are:

- Maintaining a cover of plants or residue
- Planting shelterbelts
- Strip cropping
- Increase surface roughness of soil
- Cultivating on the contour
- Maintaining soil aggregates at a size less likely to be carried by wind
- Increase stubble height
- Installing wind breaks and shelterbelts
- Irrigating the land
- Strip crop perpendicular to prevailing wind.

4.2   Remote Sensing

Crop condition monitoring is essential for crop management and yield forecasting. Traditionally, crop condition is monitored either visually or by various laboratory techniques, such as measurements of biochemical (Chlorophyll, leaf water, leaf N) and biophysical (LAI, DM) and pest/disease intensity. These techniques are and time consuming and many times may not provide a complete picture. Therefore, remote sensing data, because of its typical properties like, capability to achieve a synoptic view, potential for fast survey, capability of repetitive coverage to detect the changes, low cost involvement, higher accuracy, and use of multispectral data increased information provides a better alternative compared to the traditional methods. Remote Sensing (RS) is defined as "The science and art of obtaining the information about an object, area or phenomenon through an analysis of the data acquired by a device which is not in contact with the object". The information about the object, area and phenomenon, must be available in a form that can be impressed on a career vacuum. The information carrier or communication link is electromagnetic energy. Remote sensing data basically consists of wavelength intensity information acquired by collecting the electromagnetic radiation leaving the object at specific wavelength and measuring its intensity.

Vegetation reflectance is primarily influenced by the optical properties of plant materials viz., chlorophyll, corotinoids, water, proteins, lignin, cellulose, sugar and starch. Plant materials are composed largely of hydrogen, carbon, oxygen and nitrogen. Thus, the absorption bands observed in reflectance spectra of vegetation arise from vibrations of C-O, O-H, C-H, and N-H bonds, as well as, overtones, and combinations of these vibrations.

Infra Red Region

The absorptions from the different plant materials are similar and overlapping. So a single absorption band cannot be isolated and directly related to chemical abundance of the plant constituent. The absorption and of solar radiation is the result of many interactions with different plant materials, which varies considerably by wavelength. Water, pigments, nutrient and carbon are each expressed in the reflected optical spectrum from 400 to 2500 nm, with often overlapping, but, spectrally distint reflectance behavior. For remote sensing applications in agriculture the optical spectrum is partitioned into four distinct wavelength ranges viz., visible (400-700 nm); near infrared (700-1300 nm); shortwave infrared-1 (SWIRI-1300-1900 nm) and shortwave infrared − 2 (SWIR-II 1900-2500 nm). The typical reflectance pattern for any vegetation shows high absorption due to chlorophyll at 650 nm (red region) and high reflection due to leaf internal structure at 750 nm (NIR region) and water absorption at 950 nm and 1450 nm (SWIR region). Therefore, the whole range of 400-2500 nm is useful for agricultural purposes due to various absorption features in different regions (Table 4.1).

Table 4.1 Common spectral features for crops and soils

S.No.	Spectral region (nm)	Indication
1.	400-700	Photosynthetic pigments
2.	680	Chlorophyll absorption
3.	700-750	Chlorophyll
4.	1080-1170	Liquid water
5.	1700-1780	Various leaf waxes and Oils
6.	2100	Cellulose
7.	2100-2300	Soil properties
8.	2280-2290	Nitrogen/Protein

Thermal Region

Thermal (for infrared region) radiation refers to electromagnetic waves with wavelengths between 3.5 and 20 nm. Objects having temperature greater than absolute zero radiate energy obeying Planck's law of radiation. The windows normally used from aircraft platforms are in the 3-5 and 8-14 nm wavelength regions. Some spaceborne sensors commonly use transmission windows between 3 and 4 and between 10.5 - 12.5 nm. Remote sensing in the thermal infrared (TIR) region can be used to derive thermal indices, which independently or jointly with optical spectral indices are helpful in identifying the type of land cover and thermally characterize the objects, including vegetation. Many canopy temperature based indices have been developed for detecting plant water stress and scheduling irrigation viz., canopy-air temperature difference, stress degree days, canopy temperature variability, temperature stress day and crop water stress index.

Microwave Region

The portion of electromagnetic spectrum with wavelengths ranging from 0.3 to 100 cm are referred to microwaves. Non availability of adequate number of cloud free optical satellite datasets in the optical bio-window period is a major constraint for using optical remote sensing data for agricultural applications during monsoon season. In light of this, all weather capability of SAR, an active sensor, is an attractive option to rely upon. Microwave remote sensing techniques have all weather capability as atmosphere is transparent to microwaves at lower frequencies, penetrate clouds and are suitable for day/night operations owing to the independence of microwave sensors on sun's illumination. The radar response measured in terms of backscatter coefficient is dependent upon sensor (frequency, polarization and look angle) and target parameters (dielectric constant, surface roughness and vegetation cover).

Sensitivity of microwave radiation to surface roughness (due to soil or crop) and water content (in soil or crop) as a function of look angle and polarization at a given frequency makes SAR an attractive remote sensor for obtaining information on several crop and soil parameters. Microwave response from an agricultural field depends on both standing crop and the underlying soil conditions. Contribution to the microwave backscatter from an agricultural field is maximum from the soil during the initial stages when crop cover is negligible, mixed from soil and crop during the growth period and mostly from the crop canopy when crop cover is at its peak. Hence, to put SAR data to full use in agricultural environment, complete information of soil as well as crop conditions is essential.

Geometrical and dielectric characteristics of crops influence the interaction of microwaves and thus determine the microwave backscatter measured by the sensor. Crop phenology governs the plant water content and thus crops dielectric properties. As crops mature, the water content decreases, which in turn reduces contribution to radar backscatter from the plants. Besides crop geometry, several crop parameters such as leaf area index, plant biomass, plant water content and plant height show significant correlations with radar backscatter coefficient.

4.2.1 Geo-spatial Information and Communication Technology

The "Geo-spatial Information and Communication Technology", abbreviated as Geo-ICT is an enabling technology that is derived by integrating geo-spatial information and imaging technologies with ICT, which enables geo-computations that are effective in decision making. Geo-ICT integrates ICT with Remote Sensing (RS), Geographic Information Systems (GIS), Global Positioning System (GPS) Photogrammetry and database management systems (DBMS) in a spatio-temporal perspective. The stakeholders of the information include, farming community, bureaucrats, administrators, planners representing government for different policy level decisions/interventions, agro-industrialists, agri-business traders and academia. The developments in Geo-ICT has enabled generation of information on natural resources (relevant to agriculture) in spatio-temporal domains of different hierarchies for appropriate decisions.

4.2.2 Uses of Space Observations

The advantages of space observations emanate from several factors such as:

- Synoptic view of large areas, bringing out the inter-relations of processes of different special scales
- Frequent observations from geostationary satellites provide continuous monitoring while polar orbiting satellites give typical twice daily coverage (such data is relevant for study of weather system dynamics)
- The inherent spatial averaging is more representative than the point *in-situ* observations and readily usable for weather prediction models

- High level of uniformity of space observations overcomes the problem of inter- calibration needed for ground based instruments
- Space data covers large oceanic areas and inaccessible and remote land areas, thus giving global coverage i.e., all gaps in observations
- Parameters such as sea surface(skin) temperature, sea surface wind stress, sea level, cloud liquid water content, radiation balance, aerosol offer new types of data observations which are some of the unique parameters provided only by satellites
- Simultaneous observation of several dynamic parameters provided by different sensors in same platform facilitates study of inter-relationships and knowledge of processes (*e.g.* surface temperature and deep convection, cloud development and radiative forcing).

The agroclimatological and agrometeorological parameters are mostly defined with point observations. Sometimes, the spatial distribution of these parameters are presented through subjective approach, usually through graphics. Though innumerable computer softwares are developed for this purpose, the output is again influenced by human bias. With the advent of satellite technology, remote sensing technology became a handy in this direction. The advent of remote sensing technology and space borne sensors is a critical advance, making it possible to monitor from space a large number of earth's vital signs from atmospheric ozone to vegetation cover to sea level to glacial ice and therefore to provide a background for prediction studies. Advances in satellite instruments and their calibration and in methods of processing and archiving remotely sensed data have lead to the creation of numerous time series of environment. The two most important and directly related to agricultural studies of this nature are (a) rainfall assessment and (b) the computation of vegetative index i.e., the Normalized Difference Vegetation Index (NDVI) for crop assessment.

A. Rainfall Assessment

Global assessment of rainfall using space based measurements have been initiated long back. It is reported that the development of techniques for quantitative estimation of rainfall using data from a variety of sensors recording either the reflected solar radiation or the emitted thermal radiation from the raining clouds, has met with limited success. This is basically because all these attempts have so far dependent on measuring the radiation emanating from earth-atmosphere system in a passive mode. Active microwave probing using C and S band precipitation sensing radar appears to be promising remote sensing technique of rain from ground.

However, here also the success of quantification of rain under different ranges is quite different. That is, when the rainfall is low the accuracy is high and the rainfall is high the accuracy is low. However, one important aspect with this technique of practical use is the "Quantitative spatial gradation". This will provide a way of interpolation or extrapolation of ground based observed data in a spatial context. More work is being done to achieve success in perfect quantification of rain.

B. The Normalized Difference Vegetation Index (NDVI)

The vegetation is traditionally monitored using the information from the red (high absorption) and near infrared (high reflectance) wavelengths combined into the Normalized Difference Vegetation Index (NDVI). The NDVI is defined as:

NDVI = (NIR-RED) / (NIR+RED). Where the NIR is the reflectance in the near infrared wavelengths and RED is the reflectance in the red wavelengths.

The applicable satellite data for estimation of NDVI have so far been based on polar orbiting satellites carrying sensors detecting the radiation in red and near infrared wavelengths. Data from polar orbiting sensors have been used extensively for vegetation monitoring during the last decades based on the spectral vegetation indices, in particular NDVI. Much effort has been put into the development of improved measures of vegetation. Improvements in spectral resolution of new sensors have been in focus with American MODIS sensor. This had made estimation of new vegetation indices from MODIS sensor possible. Effective monitoring via satellite and *in situ* observations can successfully guide mitigation activities. Accurate short term forecast of NDVI could increase lead timings, making early warnings further earlier. With the launch of MODIS, unprecedented data for scientific exploration are now available to scientists in remote sensing. The MODIS sensor detects radiation in different wavebands among which two are specially suited for vegetation studies. The TERRA satellite with MODIS sensors have the great advantage over the data from the other sensors that the frequency of observation is much higher (15 minutes versus one time a day acquisitions) thus the chances to avoid cloud cover are much improved, which open up for a new vegetation monitoring scheme. One of the most important and popular methods currently used to assess crop canopy characteristics is spectral reflectance measured by using radiometers, spectrometers or digital cameras and subsequent calculation of vegetation index. However, when crop reaches certain growth stage, spectral reflectance measurements tend to reach saturation and are almost unable (less) to discriminate between canopies. However, combined output of multiple sensors would be required in future to measure both crop cover and structure. This enables the crop to be monitored throughout the growing season, which are not possible with sensing approach in isolation. This finally helps in better planting, efficient management and forecasting of final crop yield.

4.2.3 Application of Remote Sensing in Agriculture

4.2.3.1 *Crop Identification and Water Resources*

Application potentials of remote sensing techniques on resources related to some important agrometeorological services are:

- *Agricultural crops:* One basic information that remote sensing can provide to agriculture is data related to crop identification and area measurement under different types of crops or acerage estimation. This enables to estimate the total production by understanding the yield per unit area. Such information has far reaching consequences in providing adequate food security

- *Forestry and vegetation mapping:* Remote sensing can aid in providing (a) information about the extent of forest cover and give a general idea of the types of forest cover (b) forest canopy density condition (c) detection of forest hazards like fire, disease and excessive felling

- *Water resources:* Understanding water resources is important from agricultural point of view. Water supply to agriculturally related sectors depends upon the available resources, both in terms of quantity and quality. Remote sensing data is useful in assessing water resources, irrigated area studies and its monitoring and in determining potential ground water zones

- Remote sensing can also provide data related to ocean and coastal zones like identifying potential area of fish concentration, environmental degradation that takes place in coastal zones due to over exploitation

- Other promising areas of applications include disaster assessment, drought monitoring and environmental monitoring all of which have high significance.

4.2.3.2 *Monitoring Crop Condition*

Measuring and monitoring the near infrared reflectance is one way that scientists determine how healthy a particular vegetation may be. Monitoring the crop condition and yields at regional scales using imagery from operational satellites remains a challenge because of the problem in scaling local yield simulations to regional scales. Reflectance spectra have been studied intensively as a means to remotely detect the status of crops. Hyperspectral sensors measure reflectance in a large number of narrow wavebands, generally with bandwidths of less than 10 nm. With these narrow bands, reflectance and absorption features related to specific crop physical and chemical characteristics could be detected. Review of literature indicate that good relationships exist among reflectance spectra and biochemical composition, physical structure, water content and eco-physical status of plants. Numerous algorithms, ratios and indices have been developed, which attempt to detect conditions of stress within the crop. Such approaches/indices are usually normalized to minimize atmospheric interference by calculating ratios or linear combinations of two or more wavelengths within the visible and near- infrared. For these reflectance study purposes the near IR is considered from 700 to 1300 nm and the far IR from 1301 to 2550 nm. Although much hyperspectral reflectance work to date has been done at the leaf scale, typically on excised leaves, *in situ* measurements made above the canopy are becoming more widely used, driven by the need to simulate the scales involved in airborne or satellite measurements (i.e., remotely sensed imagery at the canopy scale). The ability to detect the eco-physiological status of crops and relate it to management practices at the field scale necessitates the use of remote sensing from airborne or satellite platforms. This work is intimately connected to the growing body of research in precision agriculture. In precision farming the RS techniques use ground based hyperspectral remote sensing and micrometeorological methods to monitor different plant stresses (nitrogen application rates). The results of such studies serve as an exploration in support of subsequent work at the canopy level in conjunction with aircraft- based hyperspectral measurements at the field site.

The above information (canopy-scale hyperspectral indices) is very much useful to:

- Detect both environmental and plant stresses (nitrogen stress) as they occurred throughout a growing season of field crops

- To determine if canopy-scale hyperspectral indices could be used to discriminate between management practices (fertilizer application rates) at various times throughout the growing season.

  Red edge: An important feature of spectral profile used in monitoring crop condition is the red edge (RE), which is the spectral region of the red-NIR transition (700-750 nm). This region showed large information content for vegetation spectra. The red edge region marks the boundary between the process of chlorophyll absorption in red wavelengths and within-leaf scattering in near infrared wavelengths. The slope of this region is a strong indicator of crop condition (health). When a plant is healthy with high chlorophyll content and high LAI, the red edge position shifts towards the longer wavelengths. In contrast, when it suffered from disease or chlorosis and low LAI, it shifts towards shorter wavelengths.

4.2.3.3 *Remote Sensing and N Management in Precision Agriculture*

Application Hypothesis: The color of corn crop is sensitive to nitrogen status and may provide a means to accurately match nitrogen fertilization rates to spatially variable nitrogen needs. Therefore, to find out the relationship between corn colour measured in aerial photographs and side dress nitrogen need, an experiment was conducted in USA. The results of the experiment demonstrated that corn color measured in aerial photographs could be used to predict side dress nitrogen need. Indices derived from hyperspectral reflectance spectra have the potential to be used as the indicators of environmental stress. Therefore, identifying areas of fields sensitive to weather induced stresses will allow better management of nitrogen application. Also, timing the collection of hyperspectral image data at early and late vegetative phages could enhance precision agriculture by allowing supplemental N application.

With the increased use of yield monitors on grain combines in the past two decades crop yield has repeatedly been shown to exhibit substantial spatial variation across individual fields. In addition, airborne and satellite remote sensing imagery has shown similar variation in crop growth and development throughout the growing season. Such datasets have demonstrated the potential for variable rate applications of nitrogen (N) fertilizer, based on the site-specific crop need. Applying N fertilizer site-specifically also makes sense from an environmental perspective. Several properties of the agricultural landscape that altered its susceptibility to movement and loss of nitrate-nitrogen (NO_3-N). These include soil type, topography, soil moisture, tile drainage and tillage practices. Thus, a robust method for determining appropriate N application rates must consider the spatial variability of the agricultural landscape's NO_3-N loss risk, as well as the spatial variability of the crop's N need. In this way, N prescriptions could be tailored to address both production and environmental concerns. In addition to the spatial aspects of N management, there is also a complex temporal problem that arises due to the unpredictable nature of weather patterns. Precipitation events drive the movement of NO_3-N through the agricultural system and rainfall is necessary for the crop to uptake NO_3-N from the soil. However, problems arise when precipitation events, NO_3-N availability, and crop need do not coexist in time. For example, in the mid-western United States, N is most commonly applied in the fall or spring, prior to planting corn. Nitrogen applied at these times has the greatest potential for loss to the environment, because snow melt and heavy rains in the spring season can move NO_3-N out of the agricultural system prior to crop uptake. A similar problem exists during seasons of drought. For this case, suppose NO_3-N is made available through side-dress applications of N fertilizer at mid-season, although NO_3-N is now available during the time of peak N demand, the lack of water prevents the crop from removing all the NO_3-N from the soil. The excess NO_3-N is then available for loss during precipitation events that occur after harvest, when the crop no longer needs it. Unfortunately, when making N management decisions, knowledge of future weather patterns and precipitation events is limited to the accuracy of seasonal forecasts. However, large sets of historical weather data now exist for many portions of the world, and these datasets can be used as an indicator of probable future weather patterns for an area. In this way, historical weather data becomes a useful set of information for the development of N management strategies that are conscious of the influence of weather patterns on NO_3-N movement through the agricultural system.

Over the past decade, researchers have focused on a wide variety of methods for developing site-specific N prescriptions. An arsenal of sensing techniques have been employed for identifying N deficient areas of crops, including:

- Airborne and satellite remote sensing
- Multispectral camera systems on ground vehicles
- Chlorophyll meter readings of individual corn leaves etc.

Although these sensing techniques have successfully identified N deficiencies, they do not effectively account for the spatially varying properties of the agricultural landscape or weather patterns that affect NO_3- N losses from the agricultural system. Several researchers have attempted to develop yield response functions by regressing crop yield against soil nutrient measurements, such as late-spring NO_3-N concentration and soil organic matter. High r^2 values for the relationship between crop yield and soil nutrient levels have not been consistently obtained with this approach, because the temporal aspects of N movement through the agricultural system cannot be adequately characterized by a single equation. As a result, soil nutrient concentrations based on point-in-time measurements have not been helpful for developing variable rate N recommendations. The greatest limitation in these approaches is that none of these can adequately account for N movement, which depends heavily on the temporal pattern of weather encountered during the growing season.

4.2.3.4 *Crop Growth Models and N Management*

The CERES-Maize crop growth model is another tool that has been used to study precision N management for corn (Zea mays L.). This model utilizes carbon, N, and water balance principles to simulate, in homogenous units, the daily processes that occur during plant growth and development. The final corn yield for the simulated growing season is then calculated on the harvest date. The model has been shown to adequately simulate corn growth, development, and yield on plot level, field -level, and regional scales for many locations around the world. Inputs required for model execution include (a) Management practices (plant genetics, plant population, row spacing, planting and harvest dates, and fertilizer application amounts and dates) (b) Soil factors (soil type, drained upper limit, lower limit, saturated hydraulic conductivity, root weighting factor, and effective tile drain spacing) and (c) Weather conditions (daily minimum and maximum temperature, solar radiation, and precipitation). Since, CERES-Maize utilizes N balances for crop growth analysis, it can be conveniently extended to calculate surface and subsurface NO_3-N losses. The model has undergone several modifications such that NO_3-N in run-off, tile flow and leaching can be simulated as part of the crop production process. Since CERES-Maize can collectively account for many of the spatial and temporal factors that effect crop yield and N movement through the agricultural system, it serves as a very useful and appropriate tool for developing N management strategies that address both the economic and the environmental concerns of corn production. A new decision support system called Apollo runs CERES-Maize and other DSSAT crop models for management zones within a field.

An example: Apollo is an interface that can be used to calibrate and validate model parameters and execute model runs to achieve a variety of precision farming objectives, such as prescription analysis and yield gap analysis. The Apollo system was used to calibrate CERES-Maize and run N prescriptions for an Iowa corn field divided into 100 grid cells. The overall objective was to use the results of the prescription simulations to develop a methodology for estimating the economic and environmental trade N management strategies for the corn field. The existence of a trade between the production and environmental concerns of N management is an important concept, because of the dual opposing roles that N plays in crop production and environmental quality. Whereas N fertilizer is beneficial for maximizing crop production, unused N fertilizer that is lost from the agricultural system poses a threat to environmental quality, wildlife welfare and human health. Therefore, N management strategies of the future must aim to find the appropriate balance between these opposing concerns. The first step in this endeavor is to develop a methodology for predicting how a particular N management strategy will affect corn yield and unused N remaining in the soil at harvest. With such a methodology, N management strategies can be developed and implemented with a direct understanding of the cost to the producer and the cost to the environment. In addition, the methodology could aid in the development of environmental legislation and producer compensation programs that aim to reduce the environmental risk of agricultural N management.

4.2.3.5 *Spectral Bands and N Management*

Spectral bands in the visible and near infrared regions of the spectrum have been used to develop a number of indices for estimating chlorophyll content. Reflectance in the main chlorophyll absorption region near 675 nm has been used for a long time as an indicator of chlorophyll content of leaves. However, the relationship between reflectance near 675 nm and chlorophyll content has been shown to become saturated at medium to high chlorophyll contents and consequently this index is only suitable for estimation of very low levels of leaf chlorophyll content. The commonly used normalized difference vegetation index (NDVI), defined as $(R\text{NIR}-R\text{RED})/(R\text{NIR}+R\text{RED})$, was developed by contrasting the strong chlorophyll absorption in the red wavelengths with the high reflectance in the near-infra wavelengths. The NDVI has also been found to be insensitive to medium and high chlorophyll concentrations. Recent studies have developed several new vegetation indices based on other visible wavelength bands instead of the red wavelengths near 670-680 nm.

Few researchers found good correlations between $R675/R700$ and chlorophyll (a) concentration, $R675/(R6503R700)$ and chlorophyll (b) concentration and $(R650 \times R700)$ and $R760/R500$ and total carotenoid concentration in soybean leaves. A number of other studies have found strong correlations between the reflectance around the green peak near 550 nm and chlorophyll *a* concentration. The Green NDVI is defined as

(RNIR)-(RGREEN)/(RNIR) + (RGREEN). By using the green wave length channel near 550 nm, the green NDVI is found to be more sensitive to a much wider range of chlorophyll concentrations than the original NDVI. Most of the algorithms reported in the literature have been developed using leaf reflectance measurements carried out on a few deciduous and coniferous species.

4.2.3.6 *Canopy Variables and N Management*

Measurement of various crop canopy variables during the growing season provides an opportunity for improving grain yields and quality by site-specific application of fertilizers. Important variables in this context are leaf area and total above ground biomass. Also, leaf chlorophyll and nitrogen concentration in the leaf dry matter are indicators for crop nitrogen requirements. The spatial and temporal variations in the field of these variables must be determined in order to match the crop requirements as closely as possible. Different remote sensing applications have proved to be a potential source of reflectance data for estimation of several canopy variables related to biophysical, physiological or biochemical characteristics. Hyperspectral remote sensing, or imaging spectroscopy consists of acquiring images in many (<10 nm) narrow, contiguous spectral bands, thus providing a continuous spectrum for each pixel, unlike multi-spectral systems that acquire images in a few broad spectral bands (>50 nm). Hyperspectral imaging is a powerful and versatile tool for continuous sampling and for selection of narrow wavebands, which are sensitive to specific crop variables. The possibilities for using the technique have increased over recent years, because the required spectrometers have become cheaper. Selection of new wavebands in hyperspectral imaging has been performed in a number of cases, mainly focusing on how to increase the sensitivity of the vegetation indices to chlorophyll and other pigments. These investigations have mainly been performed on leaf level, on canopies grown under controlled conditions or on vegetation very distinct from winter wheat with respect to leaf structure, canopy geometry, and background due to senescent vegetation. As a consequence of the different measurement conditions, some degree of disagreement exists in the selection of wavebands. A limited number of results have been published where hyperspectral reflectance was recorded under natural conditions in high input and high yielding crops The normalized difference vegetation index (NDVI) is the classical index, where red reflectance (Rred) and near infrared reflectance (Rnir) is used (NDVI = Rnir _Rred/ Rnir + Rred). This vegetation index has been related to crop variables such as biomass, leaf area, plant cover, leaf gap fraction, nitrogen, and chlorophyll in cereals. NDVI have during the past decades been based on either broad wavebands (50–100 nm scale) e.g., the satellite-based Landsat Thematic Mapper using the TM-spectrometer (TM), or short wavebands (10 nm scale) from field-based spectroradiometers. The broadband VIs use, in principle, average spectral information over a wide range resulting in loss of critical spectral information available in specific narrow (hyperspectral) bands. Further improvement in indices is generally obtained through the use of spectral data from distinct short bands. Some studies on hyperspectral reflectance data from field based spectrometers have been conducted on both leaf and canopy level in various plants .The results obtained in these studies indicate that knowledge of the connection between the investigated variable and the spectral data can improve the performance of vegetation indices.

Another investigation was carried out to compare the predictive power of (i) Models based on predefined short and broadbands for a normalized difference type of index (ii) The best combination of narrow wavebands for a normalized difference type of vegetation index and (iii) Partial least squares regression (PLS) using all available wavebands. The models were fitted to six different biophysical and biochemical variables; green biomass (GBM), leaf area index (LAI), leaf chlorophyll concentration (Chlconc), leaf chlorophyll density (Chl. density), leaf nitrogen concentration (Nconc) and leaf nitrogen density (N density) in winter wheat. The registrations of GBM, LAI, leaf chlorophyll and nitrogen content were performed in a plot experiment with different cultivars, plant densities and nitrogen supplies. The spectral recordings and measurements of crop variables were performed five times from early stem elongation until heading to include the variation during vegetative growth. This procedure ensured that the normally occurring variation due to canopy growth stage and management factors was included in the models, giving a more realistic basis for model development.

4.2.3.7 *Land Surface Temperature*

Land surface temperature (LST) is a key variable for environmental studies (for example, for the estimation of the fluxes at the surface/atmosphere interface). Moreover, many other applications rely on the knowledge of LST (geology, hydrology, vegetation monitoring and GCMs). In order to retrieve accurate LST values from remote sensing or satellite data, atmospheric and emissivity effects must be corrected. Several techniques are available since the 1970s for performing this correction. Most of them applied to thermal data acquired in the atmospheric window located in the region between 8 and 14 nm. Jointly with atmospheric and emissivity corrections, angular effects must also be corrected. This last effect is important for satellite sensors with a large swath angle, like MODIS (Moderate Resolution Imaging Spectroradiometer) and the NOAA (National Oceanic and Atmospheric Administration) series (with a swath angle higher than 50-), or for sensors with off-nadir view observation angles, like the ATSR (Along Track Scanning Radiometer) or AATSR (Advanced ATSR), with a forward view of 55-. In fact, the problem of the angular effects on atmospheric parameters is to a large extent solved, since radiative transfer codes like MODTRAN allow the estimation of these parameters depending on the observation angle. However, the angular variation of land surface emissivity (LSE) is not a well-known problem, especially for bare surfaces like soil or rocks. The LSE angular dependence has been studied from field and also laboratory measurements. The results obtained show lower emissivity values with increasing view angle for bare soil surfaces, whereas for dense vegetated canopies the angular dependence is minimal, in agreement with the usual assumption of Lambertian behaviour for vegetation. Some attempts have been carried out in order to parameterize in a simple way the angular variation on LSE, in which directional emissivity is given by $((h)=((0) \cos(h/2)$, where h is the view angle and $((0)$ the at-nadir emissivity. However, this expression, despite its easy application, is not appropriate for all surfaces and does not always provide good results. For sea or water surfaces, different models have been successfully developed for directional emissivity. In recent years different models have also been developed in order to analyze the angular variation over vegetation canopies, using among others the soil and vegetation emissivities as input data and the assumption of Lambertian behaviour for these components. This study addresses the simulation of the directional angular variation of emissivity. LSE is an important variable that has to be known to correct surface radiances and obtain surface temperatures. LSE is also involved in the atmospheric corrections since it appears in the reflected down welling atmospheric term. LSE analysis is very important in LST retrival. As a general result, an uncertainty on the LSE of 0.01 leads to an error on the LST of around 0.5 K. Emissivities are also important per se, so they may be diagnostic of composition, especially for the silicate minerals. LSE is thus important for studies of soil development and erosion and for estimation amounts and changes in sparse vegetation cover, in addition to bedrock mapping and resource exploration. A detailed comparison between different models can also be found in literature. Certain papers include results obtained with geometrical models as well as a discussion regarding the assumption the of Lambertian behaviour for soils.

Description of models are interesting tools because they make it possible to set up relationships between the thermal infrared (TIR) observations and surface biophysical parameters, as for example relationships between emissivity and vegetation index. Models simulate the radiance measured by a radiometer, provided that the surface, atmosphere and sensor characteristics are known. In the TIR, two major types of models can be considered: (a) Geometrical models (GM) and (b) Radiative transfer models (RTM). GM estimate the TIR radiance of a cover with the help of geometric considerations to describe the canopy structure. First, they calculate the proportions of projected surface area of the different surface components, which are directly observed in a particular view direction. GM represent the vegetation as an opaque medium and do not simulate radiative transfer with the cover. RTM estimate the cover radiance as a function of sensor viewing direction, temperature distribution and leaf angle distribution within the canopy. They simulate the propagation and the interactions within the cover of TIR radiation emitted by the cover components or incoming from the atmosphere. The canopy is represented as a set of plane elements (leaves) statistically distributed into homogeneous horizontal layers. The upward and downward radiative contributions of each layer are based upon the concept of directional gap frequency through the vegetation. The directional radiance of the cover is calculated by summing the radiative contributions of all layers. Iterations are sometimes performed to account for multiple scattering within the cover. The aforementioned models do not account for the canopy three-dimensional (3D) architecture, so they are one-dimensional (1D) models. In this respect the DART (Discrete Anisotropic Radiative Transfer) model developed by certain scientists deserves special mention, which is a 3D model and simulates the TIR radiance of vegetation covers with incomplete canopy. Moreover, other models not belonging to geometrical or radiative transfer have been developed, like for example models based on the estimation of the BRDF (Bidirectional Reflectance Distribution Function) or hybrid models.

4.3   Crop Growth Modelling

Simulation is defined as "Reproducing the essence of a system without reproducing the system itself". In simulation, the essential characteristics of the system are reproduced in a model, which is then studied in an abbreviated time scale. A model is a schematic representation of the conception of a system or an act of mimicry or a set of equations, which represents the behaviour of a system. Also, a model is "A representation of an object, system or idea in some form other than that of the entity itself". Its purpose is usually to aid in explaining, understanding or improving performance of a system. A model of an object may be an exact replica of the object or it may be an abstraction of the object's salient properties. A model is, by definition "A simplified version of a part of reality, not a one to one copy". This simplification makes models useful because it offers a comprehensive description of a problem situation. However, the simplification is, at the same time, the greatest drawback of the process. It is a difficult task to produce a comprehensible, operational representation of a part of reality, which grasps the essential elements and mechanisms of that real world system and even more demanding, when the complex systems encountered in environmental management.

4.3.1 Weather Data for Modeling

The national meteorological organizations provide weather data for crop modeling purposes through observatories across the globe. In many European countries weather records are available for over 60 years. In crop modeling the use of meteorological data has assumed a paramount importance. There is a need for high precision and accuracy of the data. The data obtained from surface observatories has proved to be excellent. It gained the confidence of the people across the globe for decades. These data are being used daily by people from all walks of life. But, the automated stations are yet to gain popularity in the under developed and developing countries. There is a huge gap between the old time surface observatories and present generation of automated stations with reference to measurement of rainfall. The principles involved in the construction and working of different sensors for measuring rainfall are not commonly followed in automated stations across the globe. As of now, solar radiation, temperature and precipitation are used as inputs in the state of art crop growth models like DSSAT.

4.3.2 Weather as in Input in Models

In crop modeling weather is used as an input. The available data is ranging from one second to one month at different sites where crop-modeling work in the world is going on. Different curve fitting techniques, interpolation and extrapolation functions are being followed to use weather data in the model operation. Agrometeorological variables are especially subject to variations in space. It is reported that, as of now, anything beyond daily data proved unworthy as they are either overestimating or under estimating the yield in simulation. Stochastic weather models can be used as random number generators whose input resembles the weather data to which they have been fit. These models are convenient and computationally fast, and are useful in a number of applications where the observed climate record is inadequate with respect to length, completeness or spatial coverage. These applications include simulation of crop growth, development and impacts of climate change. In 1995 JW Jones and Thornton described a procedure to link a third-order Markov Rainfall model to interpolated monthly mean climate surfaces. The constructed surfaces were used to generate daily weather data (rainfall and solar radiation). These are being used for purposes of system characterization and to drive a wide variety of crop and live stock production and ecosystem models.

The present generation of crop simulation models particularly DSSAT suit of models have proved their superiority over analytical, statistical, empirical and combination of two or all models so far available. In the earliest crop simulation models only photosynthesis and carbon balance were simulated. Other processes such as vegetative and reproductive development, plant water balance, micronutrients and pest and disease are not accounted for as the statistical models use correlative approach and make large area yield prediction and only final yield data are correlated with the regional mean weather variables. This approach has slowly been replaced by the present simulation models by the DSSAT models. When many inputs are added in future the models become more complex. The modelers who attempt to obtain input parameters required to add these inputs look at weather as their primary concern. They may have to adjust to the situation where they develop capsules with the scale level at which the input data on weather available.

4.3.3   Role of Weather in Decision Making

Decisions based solely upon mean climatic data are likely to be of limited use for at least two reasons. The first is concerned with definition of success and the second with averaging and time scale. In planning and analyzing agricultural systems it is essential not only to consider variability, but also to think of it in terms directly relevant to components of the system. Such analyses may be relatively straightforward probabilistic analyses of particular events, such as the start of cropping seasons in West Africa and India. The principal effects of weather on crop growth and development are well understood and are predictable.

Crop simulation models can predict responses to large variations in weather. At every point of application weather data are the most important input. The main goal of most applications of crop models is to predict commercial out-put (Grain yield, fruits, root and biomass for fodder). In general the management applications of crop simulation models can be defined as:

(a) Strategic applications (crop models are run prior to planting)

(b) Practical applications (crop models are run prior to and during crop growth) and

(c) Forecasting applications (models are run to predict yield both prior to and during crop growth).

The crop simulation models are used in USA and in Europe by farmers, private agencies and policy makers to a greater extent for decision making. Under Indian and African climatic conditions these applications have an excellent role to play. The reasons being the dependence on monsoon rains for all agricultural operations in India and the frequent dry spells and scanty rainfall in crop growing areas in Africa. Once the arrival of monsoon is delayed the policy makers and agricultural scientists in India are under tremendous pressure. They need to go for contingency plans. These models enable to evaluate alternative management strategies, quickly, effectively and at no/low cost. To account for the interaction of the management scenarios with weather conditions and the risk associated with unpredictable weather, the simulations are conducted for at least 20-30 different weather seasons or weather years. If available, the historical weather data, and if not weather generators are used presently. The assumption is that these historical data will represent the variability of the weather conditions in future. Weather also plays a key role as input for long-term crop rotation and crop sequencing simulations.

4.3.4 An Example

The following is the original work of Dr. Murthy, the author of the book. Research was conducted in Andhra Pradesh, which is one of the most important peanut growing, regions of India. The data was collected from detailed growth analysis experiments conducted during 1991 and 1994 and from yield trails from 1991 through 1997 conducted at College of

Agriculture, Hyderabad (17°34' N lat. And 78°55' long.). These data were used to calibrate and validate CROPGRO-Peanut model. The validated models were then used to simulate the yield under present climatic conditions and under future climate conditions with a range of changes in temperature, precipitation, with and without elevated CO_2 effects under irrigated and rainfed conditions. The productivities of peanut under different climate scenarios were compared with the current levels.

Effects of Temperature and Rainfall

The results of the simulation studies showed that increasing the mean temperature above the base line weather by 0.5° to 6 °C at successive intervals of 0.5 °C, at ambient CO_2 reduced yields of both irrigated and rainfed peanuts to a similar extent (Table 4.2). There was no effect of decreasing rainfall by 20% on seed yield of peanut under the present CO_2 and temperature conditions (Table 4.3). This is because the water holding capacity of the soil was good (120 mm). However, when the temperatures were increased, the reductions in pod yields linearly increased with increase in temperatures (Table 4.3).

When the rainfall was increased by 20% at ambient CO_2 and temperature conditions the pod yields were increased by about 60 kg ha^{-1}, but when temperature were also increased there was significant reductions in seed yield. However, there were greater reduction in yields when rainfall was reduced by 20% (Table 4.3). These results suggest that increase in temperature by 4.5 °C with 20% reduced rainfall can reduce peanut yields to the extent of 50%, but, if the rainfall increases by 20% the losses due to high temperatures will only be 17% (Table 4.3) indicating increased rainfall could compensate for higher temperatures to some extent.

Effects of Temperature and CO_2

Increasing CO_2 concentrations resulted in increased yields of peanut under both irrigated and rainfed conditions with 20% increase or decrease in rainfall (Table 4.2 and 4.3). This is in agreement with the results of experiments conducted in controlled environments, where peanuts grown at elevated levels of CO_2 (700 ppm) had 31% higher yields than those at ambient levels (350 ppm) at optimum temperatures. The increased yield at elevated CO_2 levels are mainly because:

(a) In C_3 species photosynthesis is limited by initial CO_2 receptor Ribulose-1-5-bisphosphate carboxylase/oxygenase (Rubisco) is substrate limited under present levels of CO_2

(b) Elevated CO_2 increases ratio of CO_2 to O_2 in the intercellular spaces, thereby promoting photosynthesis and suppresses photorespiration

(c) Partial closure of stomata induced by elevated CO_2 suppresses transpiration per unit leaf area and slows depletion of available soil moisture in water limited conditions.

However, when temperatures were also increased simultaneously with CO_2 the yields increased until 4.0°C in temperature above baseline (about 30/22 °C, day/night). But, the beneficial effects of increased CO_2 levels were nullified when the temperatures were increased $\geq$ 4.5 °C above the baseline weather under both irrigated and rainfed conditions (Table 4.3). The predicted 4.5 °C increase in the mean temperature accompanied by increased CO_2 levels will negate the beneficial effects of elevated CO_2 owing to temperature dependence of metabolic processes driving photosynthesis, fruit-set and yield. Research in controlled environments on peanut has shown that elevated CO_2 increases the productivity under irrigated and drought conditions as observed in present study. It has been shown that when daytime air temperatures reach > 33 °C (i.e. rise of 5 °C), then it reduces number of fruits and thus seed yields as observed from the seasonal analysis (Table 4.2 and 4.3).

Table 4.2 Results of sensitivity analysis for different climate change scenarios showing the simulated mean peanut seed yield (kg ha-1) at maturity and standard deviation (SD) for 25 years (1975-1999) of weather under irrigated and rainfed conditions at Hyderabad, India

Temperature increase (°C)	Irrigated				Rainfed			
	330 ppm CO_2		555 ppm CO_2		330 ppm CO_2		555 ppm CO_2	
	Yield	SD	Yield	SD	Yield	SD	Yield	SD
0.0	1570	112	2223	144	1431	219	2041	297
1.0	1526	120	2168	157	1369	234	1977	319
1.5	1484	119	2127	153	1322	224	1926	318
2.0	1451	130	2100	173	1294	236	1891	330
2.5	1436	132	2095	178	1274	240	1876	334

Table 4.2 *contd...*

Temperature increase (°C)	Irrigated				Rainfed			
	330 ppm CO_2		555 ppm CO_2		330 ppm		555 ppm CO_2	
	Yield	SD	Yield	SD	Yield	SD	Yield	SD
4.0	1210	551	1754	797	1060	480	1560	699
4.5	741	735	1077	1064	770	589	1148	869
5.0	377	621	555	916	530	560	662	818

Table 4.3 Results of sensitivity analysis for different climate change scenarios showing the simulated mean peanut seed yield (kg ha^{-1}) at maturity and standard deviation (SD) for 25 years (1975-1999) of weather under rainfed conditions at Hyderabad, India with 20 % increase or decrease in rainfall

Temperature increase (°C)	Rainfed (+20% rainfall)				Rainfed (-20% Rainfall)			
	330 ppm CO_2		555 ppm CO_2		330 ppm CO_2		555 ppm CO_2	
	Yield	SD	Yield	SD	Yield	SD	Yield	SD
0.0	1494	106	2140	129	1430	219	1948	354
1.0	1437	95	2060	117	1298	294	1873	380
1.5	1400	88	2018	114	1254	283	1821	378
2.0	1368	82	1990	110	1222	295	1780	394
2.5	1356	82	1980	122	1195	295	1762	416
3.0	1367	93	2005	135	1177	315	1742	440
3.5	1385	106	2035	157	1095	394	1628	556
4.0	1346	301	1971	444	1015	452	1518	646
4.5	1238	475	1795	693	700	563	1048	830
5.0	1000	640	1213	1016	416	493	647	764

5

Climate Change – Agriculture

*It may be that I shall find it good to get outside my body to cast it off
like a worn-out garment. But, I shall not cease to work. I shall inspire
men everywhere, until the world shall know that it is the one with God.*

- Swamy Vivekananda

5.1   Climate Change

One of the major challenges the human kind facing is to provide an
equitable standard of living for the current and future generations:
adequate food, water, energy, safety, shelter and a healthy environment.
Human induced climate change and increasing climate variability as well
as other global environmental issues such as land degradation, loss of
biological diversity, increasing pollution of the atmosphere and fresh
water and stratospheric ozone depletion threaten human ability to meet
these needs. Therefore, the science of climate change and variability and
their likely impacts on agriculture assumed significance.

Definitions

Climate Change: Climate change is defined as "A change in the state of
the climate that can be identified by changes in the mean and or the
variability of its properties and that persists for an extended period
typically decades or longer".

The climate change may be due to

- Natural internal processes
- External forcings
- Persistent anthropogenic changes.

Any one or combination of the above processes cause changes in the
composition of atmosphere or land use.

Climate variability: Climate variability is defined as "Variations in the mean state and other statistics of the climate on all temporal and spatial scales beyond that of individual weather events".

The climate variability may be due to

- Natural internal processes within the climate system (internal variability)
- Variations in natural or anthropogenic external forcings (external variability).

5.1.1 Climate Change - Global Processes and Effects

I. Human activities

A. Land Use Change

    (a)   Urbanization

    (b)   Deforestation

    (c)   Land conversion to agriculture

    (d)   Increase in impermeable surface.

B. Fossil fuel burning

(a) Industry

- Chemicals
- Cement

(b) Energy production

- Heating
- Electricity
- Power plants

(c) Transport

- Plane traffic
- Shipping freight
- Trucking freight
- Cars

(d) Agriculture

- Fertilizers
- Chemicals.

II.   Climate Change Processes

A.   Enhanced Greenhouse effect

- Carbon cycle disturbances
- CO_2
- CH_4
- N_2O

B.   Global Warming (average temperature rise)

(a) Ice caps melting (Sea level rise)

(b) Precipitation changes

(c) Cloud cover changes

(d) Ocean circulation upheaval

- Salinity
- Water temperature

(e) Gulf stream modification

(f) Monsoon disturbances.

III.   Disasters

A. Environmental refugees

(a) Droughts

(b) Cyclones

(c) Floods

(d) Tsunami

(e) Wild fire.

B.   Biodiversity losses

- Coastal wetlands disappearing
- Coral bleaching

C.   Casualties

D.   Economic losses

E.   Diseases spread

- Infectious diseases (vector change)
- Diarrohoea
- Cardio-respiratory diseases

F.   Subsistence farming and fishing at stake

    (a) Malnutrition

    (b) Traditional lifestyles

    (c) Coastal wetlands disappearing.

5.1.2   Earth's Climate

The climate is a complex, interactive system consisting of the atmosphere, land surface, snow, ice, oceans, other bodies of water and living things. The atmospheric component of the climate system most obviously characterizes climate. Climate is usually described in terms of the mean and variability of temperature, precipitation and wind over a period of time, ranging from months to millions of years (the classical period is 30 years). The climate system evolves under the influence of its own internal dynamics and due to changes in external factors that affect climate (called "forcings"). Solar radiation powers the climate system.

As rising concentrations of gases warm earth's climate snow and ice begin to melt. This melting reveals darker land and water surfaces that were beneath the snow and ice and these darker surfaces absorb more of sun's heat, causing more warming which causes more melting and so on in a self reinforcing cycle. This feedback loop known as "Ice-albedo-feedback" amplifies the initial warming caused by rising levels of greenhouse gases.

5.1.3 Relationship between Climate Change and Weather

Climate change and weather are intertwined (average weather is climate). Observations show that there have been changes in weather and it is the statistics of changes in weather over time that identify climate change. While weather and climate are closely related, there are important differences. A common confusion between weather and climate arises when scientists are asked how they can predict climate fifty years from now when they cannot predict the weather a few weeks from now. The chaotic nature of weather makes it unpredictable beyond a few days. Projecting changes in climate (i.e., long term average weather) due to changes in atmospheric composition or other factors is very different and much more manageable issue.

While many factors continue to influence climate, scientists have determined that human activities have become a dominant force and are responsible for most of the warming observed over the past fifty years. Human caused climate change has resulted not only from changes in the amounts of greenhouse gases in the atmosphere, but also from changes in small particles (aerosols), as well as from changes in land use. As climate changes, the probabilities of certain types of weather events are affected. For example, as earth's average temperature has increased, some weather phenomena have become more frequent and intense (e.g., heat waves and heavy downpours), while others have become less frequent and intense (e.g., extreme cold events). Climate can be viewed as concerning the status of the entire earth system that serve as the global background conditions that determine weather patterns. For example, an "EL NINO" effecting the weather in coastal Peru limits on the probable evolution of weather patterns that random effects can produce. A "La Nina" would set different limits.

5.1.4 The Greenhouse Effect

The sun powers earth's climate, radiating energy at very short wavelengths, predominately in the visible or near-visible (*e.g.*, ultraviolet) part of the spectrum. Roughly one-third of the solar energy that reaches the top of earth's atmosphere is reflected directly back to space. The remaining two-thirds is absorbed by the surface and to a lesser extent, by the atmosphere. To balance the absorbed incoming energy, the earth must, on average, radiate the same amount of energy back to space. Because the earth is much colder than the sun, it radiates at much longer wavelengths, primarily in the infrared part of the spectrum. Much of this thermal radiation emitted by the land and ocean is absorbed by the atmosphere, including clouds, and reradiated back to earth. This is called the "Greenhouse effect". The glass walls in a greenhouse reduce airflow and increase the temperature of the air inside. Analogously, but through a different physical process, the earth's greenhouse effect warms the surface of the planet. Without the natural greenhouse effect, the average temperature at earth's surface would be below the freezing point of water. Thus, earth's natural greenhouse effect makes life possible. However, some of the human activities, are causing global warming.

The two most abundant gases in the atmosphere, nitrogen (comprising 78% of the dry atmosphere) and oxygen (comprising 21%), exert almost no greenhouse effect. Instead, the greenhouse effect comes from molecules that are more complex and much less common. Water vapour is the most important greenhouse gas and carbon dioxide (CO_2) is the second-most important one. Methane, nitrous oxide, ozone and several other gases present in the atmosphere in small amounts also contribute to the greenhouse effect.

Several components of the climate system, notably the oceans and living things, affect atmospheric concentrations of green-house gases. A prime example of this is plants taking CO_2 out of the atmosphere and converting it (and water) into carbohydrates via photosynthesis. In the industrial era, human activities have added greenhouse gases to the atmosphere, primarily through the burning of fossil fuels and clearing of forests. Adding more of a greenhouse gas, such as CO_2 to the atmosphere intensifies the greenhouse effect, thus warming earth's climate.

5.1.5 Temperature Changes on the Earth

Instrumental observations over the past 157 years show that temperatures at the surface have risen globally, with important regional variations, expressed as global average surface temperatures have increased by 0.74 °C between 1906 and 2005. Warming in the last century has occurred in two phases, from the 1910s to the 1940s (0.35 °C), and more strongly from the 1970s to the present (0.55 °C). An increasing rate of warming has taken place over the last 25 years and 11 of the 12 warmest years (1995-2006) on record have occurred in the past 12 years. Above the surface, global observations since the late 1950s show that the troposphere (up to about 10 km) has warmed at a slightly greater rate than the surface, while the stratosphere (about 10-30 km) has cooled markedly since 1979. This is in accord with physical expectations and most model results.   Confirmation of global warming comes from warming of the oceans, rising sea levels, glaciers melting, sea ice retreating in the Arctic and diminished snow cover in the Northern Hemisphere.

5.1.6 Precipitation Changes on Earth

Observations show that changes are occurring in the amount, intensity, frequency and type of precipitation. These aspects of precipitation generally exhibit large natural variability, and El Nino and changes in atmospheric circulation patterns such as the North Atlantic Oscillation have a substantial influence. Pronounced long-term trends from 1900 to 2005 have been observed in precipitation amount in some places: significantly wetter in eastern north and south America, northern Europe and northern and central Asia, but drier in the Sahel, southern Africa, the Mediterranean and southern Asia. More precipitation now falls as rain rather than snow in northern regions. Widespread increases in heavy precipitation events have been observed, even in places where total amounts have decreased. These changes are associated with increased water vapour in the atmosphere arising from the warming of the world's oceans, especially at lower latitudes. There are also increases in some regions in the occurrences of both droughts and floods.

As the temperatures rise, the likelihood of precipitation falling as rain rather than snow increases, especially in autumn and spring at the beginning and end of snow season and areas where temperatures are near freezing. Such changes are observed in many places, especially over land and middle and high latitudes of the Northern Hemisphere, leading to increased rains but reduced snow packs and consequently diminished water resources in summer.

5.1.7 Changes in Extreme Events

Since 1950, the number of heat waves increased and widespread increases have occurred in the numbers of warm nights. The extent of regions affected by droughts has also increased as precipitation over land has marginally decreased while evaporation has increased due to warmer conditions. Generally, heavy daily precipitation events that lead to flooding have increased. Tropical storm and hurricane frequencies vary considerably from year to year, but, evidence suggests substantial increases in intensity and duration since the 1970s. In the extra-tropics, variations in tracks and intensity of storms reflect variations in major features of the atmospheric circulation, such as the North Atlantic Oscillation.

A prominent indication of a change in extremes is the observed evidence of increases in heavy precipitation events over the mid latitudes in the last 50 years, even in places where mean precipitation amounts are not increasing. For heavy precipitation events, increasing trends are reported as well.

5.1.8 Sea Level Rising

There is a strong evidence that global sea level gradually rose in the 20th century and is currently rising at an increased rate, after a period of little change between AD 0 and AD 1900. Sea level is projected to rise at an even greater rate in this century. The two major causes of global sea level rise are thermal expansion of the oceans (water expands as it warms) and the loss of land based ice due to increased melting. Satellite observations available since the early 1990s provide more accurate sea level data with nearly global coverage. This decade long satellite altimetry data set shows that since 1993, sea level has been rising at a rate of around 3 mm yr^{-1}, significantly higher than the average during the previous half century. Coastal tide gauge measurements confirm this observation. Global sea level is projected to rise during the 21st century at a greater rate than during 1961 to 2003. Global sea level reaches 0.22 to 0.44 m above 1990 levels and is rising at about 4 mm yr^{-1}. The thermal expansion is projected to contribute more than half of the average rise, but, land ice will lose mass rapidly as the century progresses.

5.1.9   Tackling Climate Change

Climate change may not be a smooth linear process of a world warming gradually and steadily, but rather a series of sudden jolts, like flips from one stable climate to another, radically different. The pace of change is already much faster than expected 10 years ago. Without drastic change, its impacts appear certain.

Climate change and its effects mater fundamentally to everyone. What is at stake is not comfort, but survival. Food security, political security etc., are under stake. Scientists have consistently been trying to get this message across in every way possible, including the use of media. Gradually, the efforts to disseminate the warnings of science are beginning to pay off. This is the start of the change the planet needs.

Climate change mitigation is trying to reduce the expected impact. This includes new policies, innovative technologies and change in day to day activities. Climate change adaptation is preparing to cope with inevitable changes ahead (higher temperature, scarcer water resources, more frequent storms etc). Adaptation aims at "Reducing the risk and damage from current and future harmful impacts cost effectively". Mitigation and adaptation can compliment each other and together can significantly reduce the consequences of anthropogenic climate change (change caused by human activities).

Green house gases have both natural and manmade sources. There are many natural processes that release and store GHGs like volcanic activity and swamps which account for considerable amounts of GHG emissions. Their concentration in the atmosphere consequently also varied in pre-industrial times. But, today atmospheric concentrations of CO_2 and CH_4

far exceed the natural range 6,50,000 years ago. These enormous amounts of GHG are closely linked to human activities, such as fossil fuel combustion and land use change that release GHGs into the atmosphere. Nature is not capable of balancing this development.

Individual responsibility for climate change mitigation decreases with decreasing economic power. In poor countries more responsibility lies with those who can act, such as governments and companies. In order to stay below global 2 °C temperature rise, the UN development programmes 2008 (human development report) suggested emissions reductions by developed countries of 80% by 2050, with 30% reductions by 2020. Under this scenario, developing countries would need to cut their emissions by 20% by 2050, with emissions rising until 2020. Average emissions in both developed and developing countries would converge by 2060 to about 2.0 tons per head of CO_2 equivalent.

5.1.9.1 *Tackling Climate Change Mitigation*

Climate change mitigation is defined as "Any action taken to permanently eliminate or reduce the long term risks and hazards of climate change to human life and property". According to IPCC "An anthropogenic intervention to reduce the sources or enhance risks of green house gases".

The key to success for an effective emissions reduction programme is to have a well organised, performing structure and a clear process in place. The following "all in one" strategies are sorted by cost efficiency for mitigation.

Savings

  (a) Standby losses

  (b) Use of sugarcane bio-fuels

  (c) Fuel efficient commercial and other vehicles

  (d) Water heating

  (e) Air conditioning

  (f) Lighting systems

  (g) Fuel efficient commercial vehicles.

Costs

  (a) Nuclear

(b) Small transit

(c) Small hydro

(d) Industrial non CO_2

(e) Airplane efficiency

(f) Forestation

(g) Coal to gas shift

(h) Biodiesel.

5.1.9.2   *Tackling Climate change Adaptation*

Climate change adaptation is defined as "The ability of a system to adjust to climate change (including climate variability and extremes) to moderate potential damage, to take advantage of opportunities or to cope with consequences". The IPCC defined adaptation as "The adjustment in natural or human systems to a new changing environment." The various types of adaptation are:

(a) Anticipatory and reactive

(b) Private and public

(c) Autonomous and planned.

Adaptation actions are taken to cope with a changing climate, e.g., increasing rainfall, higher temperatures, scarcer water resources, more frequent storms which are occurring at present or anticipating such changes in future. Examples of actions

(a) Using scarce water more efficiently

(b) Adapting existing building codes to withstand future climate conditions and extreme weather events

(c) Construction of flood walls and raising levels of dykes against sea level rise

(d) Use of public transport

(e) Use of solar, wind and ocean energies

(f) Efficient waste management.

5.1.9.3   *Tackling Climate change as Individuals*

Adaptation and mitigation are two types of policy response to climate change, which can be complimentary, substitutable or independent of each other. However, in poor countries more responsibility lies with those who can act, such as governments and companies. Everybody starts somewhere and has room for improvement.

As a consumer

(a) Buy high quality

(b) Choose seasonal products

(c) Choose local products

(d) Try organic products

(e) Drink tap water

(f) Choose products with limited packaging.

As a resident

(a) Turn off electric devices when not using them (make sure that they do not remain in standby mode)

(b) Turn off the light when leaving a room

(c) Collect the rain water for gardening

(d) Choose low energy bulbs

(e) Put a lid on pans when boiling water

(f) Run washing machines during slack hours

(g) Improve insulation (window, roof walls)

(h) Replace very old electric devices

(i) Use water saving tap inserts

(j) Use water saving shower heads

(k) Choose collective instead of individual building

(l) Choose ecological material, locally extracted and manufactured

(m) Choose renewable energies.

As a mover

(a) Limit flying

(b) Limit car use

(c) Replace very old cars

(d) Limit your speed

(e) Drive with fluidity

(f) Respect pedestrians and bikers

(g) Use your car only if no other option

(h) Bike or walk

(i) Use video conferencing instead of travel

(j) Use public transportation.

As a citizen

(a) Spread the word of climate change

(b) Get involved.

As a parent

(a) Educate your children to save energy

(b) Educate your children to use resources when needed.

As a professional

(a) Invest in renewable energies

(b) Invest in low energy sectors

(c) Turn off computers when leaving

(d) Switch off printers and copy machines at night (make sure they do not remain in standby mode)

(e) Print only when necessary.

At national/ international scale

(a) Set performance standards of power plants, gas processing plants, refineries, chemical industries, cement factories, iron and steel manufacturing

(b) For transport expand the national public transport, develop rail and river freight alternatives

(c) In agriculture limit/ control fertilizers and pesticides, preserve biological sinks like forests, follow organic farming where ever possible

(d) In energy use prefer solar, wind and ocean

(e) Organize awareness camaigns and educational campaigns

(f) In construction promote local and ecological building, set public buildings as an example

(g) In waste management support eco design projects, organize sorting and recycling of waste

(h) In urban planning limit urban sprawl, control and limit the use of cars in city limits

(i) In public transport expand the public transport network, run a reliable and regular service, make it affordable, make it easy for everyone to use.

5.2   Impacts of Climate Change on Agriculture

Technological developments have allowed remarkable progress in agricultural output per unit of land, increasing per capita food availability despite a consistent decline in per capita agricultural land area. Production of food and fibre has more than kept pace with the sharp increase in demand in a more populated world, so that the global average daily availability of calories per capita has increased, though with regional exceptions. The absolute area of global arable land has increased to about 1400 mha, an overall increase of 8% since the 1960s (5% decrease in developed countries and 22% increase in developing countries). This trend is expected to continue into the future, with a projected additional 500 mha converted to agriculture from 1997-2020, mostly in Latin America and sub-saharan Africa.

Emission Trends

Agriculture accounts for an estimated emission of 10-12% of total global atmospheric emissions.

Of global anthropogenic emissions in 2005, agriculture accounted for about 60% of N_2O and about 50% CH_4.

Without additional policies, agricultural N_2O and CH_4 emissions are projected to increase by 50% and 60% respectively by 2030.

Climate change and variability, drought and other climate related extremes have a direct influence on the quantity and quality of agricultural production and in many cases, adversely affect it especially in developing countries, where technology generation, innovation and adoption are too slow to counteract the adverse effects of varying and changing environmental conditions. The interdisciplinary nature of these issues requires a long lasting and where possible more substantial role for agro meteorology in the efforts to promote sustainable agricultural development during the 21st century. There is a need to develop locally new and better agro-meteorological adaptation strategies to increasing climate variability and climate change, especially in vulnerable regions where food production is most sensitive to climatic fluctuations.

5.2.1   Tropical Regions

Observed climate in India

- India's annual mean temperature showed significant trend of 0.56 °C increasing per 100 years during the period 1901-2007

- The all India maximum temperature shows an increase in temperature by 1.02 °C per 100 years
- In addition to natural trends of increasing temperatures from June-May every year, the trend in daily maximum temperature in India is also observed to be increasing from January, attaining a peak in the month of May
- Beyond May, the temperature starts decreasing up to December. This trend is also in addition to natural decreasing trend
- The all India mean annual minimum temperature has significantly increased by 0.12 °C per 100 years during the period 1901-2007
- The all Indian monsoon rainfall series (based on 1871-2009 data) indicates that the mean rainfall is 848 mm per year, with a standard deviation of 83 mm
- It is seen in 139 year period that there are total 23 deficient years and 20 excess years and the remaining are normal years
- It is observed that the excess and deficit years are more frequent
- Extreme rainfall and its intensity is increasing at many places in India.

Projected climate in India

- There may not be significant decrease in the monsoon rainfall in the future
- The rainy days appear to be less in number in future than the present
- Increase in rainfall intensity in the 21st century is projected over most of the regions
- Daily extremes in surface air temperature may intensify in future
- The annual mean surface air temperature rise by the end of the 21st century ranges from 3.5 °C to 4.3 °C
- Rise of more than 4.5 °C in night time temperature may be seen throughout India
- Roughly 40% of the world population lives in the tropics and agriculture is a very important sector for the economies of most countries in the tropics. In tropical Asia, more than half of the labour force is employed in agriculture, accounting for 10-63% of the GDP in most countries of the region
- The arid and semiarid regions account for approximately 30%, of the world total area and are inhabited by 1.1 billion people

- In India, majority of river systems have shown increase in precipitation at the basic level.

- A rise in atmospheric CO_2 to 550 ppm under controlled environmental conditions enhanced the yields of wheat, chickpea, greengram, pigeonpea, soybean, tomato and potato from 14% to 27%

- In coconut, arecanut and cocoa increased CO_2 lead to higher biomass

- In hybrid and its parental lines in rice, elevated CO_2 positively affected grain quality traits but negatively affected traits like aroma, protein and micronutrient contents

- Sunflower hybrids grown under elevated CO_2 showed a significant increase in biomass (61% to 68%) and grain yield (36% to 70%)

- An increase in 1 °C in mean temperature associated with CO_2 increase, would not cause any significant loss in wheat if planting dates and varieties are changed

- Cotton yield may decrease in northern India as compared to south India

- Total potato production and productivity varies among different agroecological zones

- Rainfall may influence the incidence of downey mildew disease in grapes during the months of February, March and April, when berries mature

- Increase in minimum temperature during fruit maturity affect the quality in grapes

- Temperature change will benefit apple cultivation in high altitude regions

- The agricultural climate of the arid and semiarid tropical regions in Asia, Africa and Latin America, add additional layers of risk and uncertainty to agricultural systems that are already affected by land degradation due to growing population pressures

- The impacts of climate change in Africa are likely to be more than in other regions. An increase of 5-8% (60-90 million ha) of arid and semi-arid land in Africa is projected by 2080s

- Under a range of climate change scenarios it was found that declining agricultural yields are likely due to droughts and land degradation; Ecosystems are likely to experience major shifts in species

- In certain agroecological zones such as the southern Sahelian zone of west Africa, where the predominant soils are sandy in nature, surface soil temperatures could exceed even 60 °C and that under such temperatures, enzymes degradation will limit photosynthesis and growth

- The dominant impact of global warming is predicted to be a reduction in soil moisture in subhumid zones and a reduction in runoff

- The general conclusion is that climate change will affect some parts of Africa negatively, although it will enhance prospects for crop production in other areas. Agroecological suitability in the highlands of Kenya would increase by 20% with warming of 2.5 °C based on an index of potential food production. In contrast, semiarid areas are likely to be worse off. In eastern Kenya, 2.5 °C of warming results in a 20% decrease in calorie production

- The predicted increase in frequency and/or severity of extreme events coupled with any increase in intensity of tropical cyclones could further exacerbate adverse impacts of climate change on the agricultural sector in Asia

- Studies under elevated CO_2 conditions, suggest that, in general, areas in mid and high latitudes will experience increases in crop yield, whereas yields in areas in the lower latitudes will decrease. Generally climatic variability and change will seriously endanger sustained agricultural production in tropical Asia in coming decades

- The scheduling of the cropping season as well as the duration of the growing period of the crops would also be affected

- Studies conducted in India, Indonesia and the Philippines confirmed that spikelet sterility and reduced yields negate any increase in dry-matter production as a result of CO_2 fertilisation

- Rice yields in east Java could decline by 1% annually as a result of increases in temperature. Studies in Asia revealed adverse effects on sorghum in rainfed areas of India, for corn yields in the Philippines on rice and on the tea industry of Sri Lanka

- Rise in sea level by 1 m would lead to loss of mangroves

- Coastal areas in Asia will be at risk due to increased flooding

- Crop yields could increase up to 20% in east and south east Asia, while they could decrease up to 30% in central and south Asia by mid 21st century

- Agricultural irrigation demand in arid and semi-arid regions of east Asia is expected to increase by 10% for an increase in temperature of 1 °C

- Agricultural production in lower latitude and lower income countries is more likely to be negatively affected by climate change

- A good number of researchers have all concluded that climate change would affect agriculture as a result of increased temperatures, changes in rainfall patterns and increased frequency of extreme events, which could cause changes in pest ecology, ecological disruption in agricultural areas and socioeconomic shifts in land use practices

- Reductions in rainfall in arid and semi-arid regions are likely to lead to water shortages in Latin America

- Increases of 2 °C and decreases in soil water would lead to replacement of tropical forest

- Cattle productivity in Latin America is likely to decline in response to 4 °C increase in temperatures

- Extremes in climate variability already severely affected agriculture in Latin America. The largest area with marked vulnerability to climate variability in Latin America is northeast Brazil. Periodic occurrences of severe E1 Nino-associated droughts in northeastern Brazil have resulted in occasional famines. Under doubled CO_2 scenarios, yields are projected to fall by 17 to 53% depending on whether direct effects of CO_2 are considered

- In Africa, mountain environments are potentially vulnerable to the impacts of global warming, and its important ramifications include mountain streams, water management, agriculture and tourism

- Production from agriculture and forestry by 2030 is projected to decline in Australia and New Zealand

- Growth rates of economically important plantation crops (for example: Pinus radiate) are likely to increase with CO_2 fertilisation, warmer winters and wetter conditions

- The survival rate of pathogens in winter or summer could vary with an increase in surface temperature. Higher temperatures in winter will not only result in higher pathogen survival rates but also lead to extension of cropping area, which could provide more host plants for pathogens

- Damage from disease may be more serious because heat stress conditions will weaken the disease resistance of host plants and provide pathogenic bacteria with more favourable growth conditions

- Climate affects livestock in four ways: Through
    (i)   The impact of changes on availability and price of feed grain
    (ii) Impacts on livestock pastures and forage crops
    (iii) The direct effects of weather and extreme events on animal health, growth and reproduction
    (iv) Changes in the distribution of livestock diseases.

- Climate change has potential to exacerbate the loss of biodiversity in semi arid regions

- The impact of climate variability on livestock is generally negative in the humid and sub-humid tropics, particularly in the latter. For animals, heat stress has a variety of detrimental effects with significant effects on milk production and reproduction in dairy cows and swine fertility

- Livestock in humid areas in Africa are prone to disease such as those carried by the tsetse fly. With warming, its distribution could extend westward in Angola and northeast in Tanzania but with reductions in the prevalence of tsetse in some current areas of distribution.

5.2.2 Temperate Regions

- Increased climate variability has resulted in greater fluctuations in crop yields during recent decades in temperate regions

- Extreme weather events such as drought, flooding and heat waves have had severe impacts on agriculture and forestry, as have changes in drought tendencies, soil moisture availability and frost free growing seasons

- Agriculture has also played a role in greenhouse gas emissions. The clearing of forests, the draining of wetlands and the ploughing of rangelands have led to a significant increase in atmospheric CO_2, as organic carbon was decomposed

- Nitrous oxide (N_2O) originates as a byproduct of nitrogen fertilizer application and in water logged soils. Thus, in higher latitudes, a spring burst of N_2O emissions occur with rapid snowmelt. Heavy rains in low lying areas also cause a N_2O burst of emissions. Methane (CH_4) emitted from agriculture is produced by the microbial breakdown of plant material and in the digestive system of cattle

- In Europe winter precipitation exceeds above normal

- Annual run off is projected to increase in northern Europe and decrease in Southern Europe

- Flora could become vulnerable and by 2050 crops are expected to show a northward expansion in area

- Forested area is likely to increase in north and decrease in the south of Europe

- The combined effect of climate change and enhanced CO_2 on crop production varies. Yields of C_3 crops (vegetables, wheat, grapes) generally increase. Yields of C_4 crops (corn, sugarcane, tropical grasses) generally decrease. However, annual variability of crop yields increase

- Distinct regional patterns by latitude were discernible in future climate scenarios for Europe and North America. Temperatures are expected to increase in nearly all areas but the largest temperature increases are projected over southern portions of both the United States and Europe

- The extreme cold of winter is expected to diminish but a greater likelihood of heat waves is projected in summer

- An increase in the frequency and intensity of heavy precipitation is expected, even in southern Europe and the southern United states, despite projections of total precipitation to decrease

- Northern crop areas of both Europe and the United states will have a longer growing season and an expansion of suitable area for crop production

- With higher crop production, however, the increased risk of nutrient leaching and an accelerated breakdown of soil organic matter may affect the quality of northern agricultural lands

- Lower crop yields are expected in southern crop areas due to the warmer and drier summers

- Sea level rise, coastal vulnerability, severe heat waves and by mid century daily average ozone levels are projected to increase in north America
- Moderate climate change in early decades of the century is projected to increase aggregate yields of rainfed agriculture by 5-20% in temperate north America
- Forests are likely to be effected through fires, pests and diseases.

5.2.3   Tackling Climate Change in Agriculture

5.2.3.1 *Adaption in Agriculture*

Whether or not there will be significant climate change, inherent climatic variability makes adaptation unavoidable. These are embedded on issues such as sustainability of land productivity, changes in erosion, degradation and environmental quality, which also require due consideration.

Adaptation in Tropical Regions

Earlier planting dates or cultivar substitution and methods of microclimatic modification

- Development of physiological based animal models with well developed climate components
- Improvement of carbon sequestration is required from agriculture through permanent land cover, utilizing conservation tillage, reducing fallow land in summer, incorporating rotations of forage and improving nutrient management with fertilizers
- Standardisation of crop models and monitoring of crop development and growth together with appropriate climate information
- Improved management practices and development of water conservation strategies, both from traditional and modern practices
- Planting of more shelterbelts or the use of scattered trees amongst crops for the reduction of erosion and wind damage and conservation of moisture
- Implementation of sustainable agriculture and forestry practices will both conserve land and improve yields over the long-term
- Development of innovative new technologies (e.g., climate forecasting) alongside traditional methods (e.g., intercropping, mulching) will be needed for yield improvement

- Development of adaptation strategies such as response farming at the local community level will engage active participation of the land users.

Adaptation in Temperate Regions

- Earlier planting with the use of long season varieties to increase crop yields
- In hotter regions, the introduction of shorter season varieties
- Stabilization of production and conservation of soil moisture
- Use of shorter crop rotations and routine crop thinning in areas that experience higher precipitation
- Larger spacing in forestry plantation and later thinning to reduce impacts of drought
- Increased the application of integrated pest management techniques.

5.2.3.2   *Mitigation in Agriculture*

Mitigation measures are most essential for increasing production and productivity in agriculture.

Mitigation in tropical regions

- Reduced tillage intensity and summer follow areas
- Improved manure management and feed rotations
- Integrated nutrient and pest management
- Restoration of degraded lands, erosion control, organic amendment and nutrient amendments
- Rice crop management to reduce CH_4 emissions (Example ; SRI cultivation)
- Improved energy efficiency in agriculture and using bio-energy depending on relative prices of fuel
- Substitution of fossil fuels by energy production from agricultural feed stocks, viz., crop residues, dung and energy crops.

Mitigation in Temperate Regions

- Prevention of over grazing of grasslands, fire management and species introduction
- Allocation of summer follow areas for pasture and rangeland agriculture

- Introduction of forage cropping into rangeland and pasture rotations
- Restoration of cultivated organic soils
- Improved livestock management through improved feeding practices, dietary additives, improved manure management, improved anaerobic digestion, breeding and other structural changes
- Afforestation, reforestation, forest management and reduced deforestation
- Harvested wood product management
- Use of forest products for bio-energy to replace fossil fuel use.

5.3   Climate Neutral

Climate change is the defining issue of 21st century. The issue stays until it is appropriately addressed. Climate neutral future is one of the major options. It takes patience, persistence and determination and it can be done.

A way which produces no net greenhouse gas (GHG) emissions which could be achieved by reducing greenhouse gas emissions as much as possible and using carbon offsets to neutralize the remaining emissions.

The journey to climate neutrality is not a straight line, but, a cycle a matter of cutting down greenhouse gas emissions.

5.3.1   Reasons for Climate Neutral

There are four reasons for climate neutral and reducing climate footprint.

Sparing the Climate

The buildup of GHGs threatens to set the earth on the path to a predictably different climate. The IPCC estimated that many parts of the planet will be warmer. Droughts, floods, another form of extreme weather will become more frequent, threatening food supplies. Plants and animals which cannot adjust may face consequences. Sea levels are rising and will continue to do so, forcing hundreds of thousands of people in coastal zones to migrate. One of the main GHGs which humans are adding to the atmosphere, carbon dioxide is rapidly increasing. Around 1750 AD, about the start of the industrial revolution in Europe, there were 280 ppm of CO_2 in the atmosphere. By 2009 the overall amount of GHGs has topped 390 ppm. Scientists emphasize that the earth's average temperature should not rise by more than 2 °C over pre-industrial levels.

Among others, European Union indicated that it is essential to minimize the risk of what the UN framework convention for climate change calls dangerous climate change and keep the costs of adapting to a warmer world bearable. A 50 percent chance of keeping the 2 °C if the GHG concentration remains below 450 PPM.

Conserving Natural Resources

There is a growing evidence of another and quite different threat developing. Humans may soon run short of the fossil fuels (gas and oil) which keep modern society going due to natural deflection. By 2030, world energy use will probably have increased by more than 50%. Earth can attain energy security when it moves from fossil fuels to fossil free alternatives

Protecting Human Health

Emissions linked to burning of fossil fuels (SO_2, NO_2 etc) often make people ill. Three million people die every year due to outdoor air pollution. Pollution comes from vehicles, power stations and factories. It also damages natural world, through acid rain and smog.

Boosting Economy

Individuals who reduce their energy consumption and its climate impact also save money. On a more macroeconomic level, economic opportunities arise from measures taken to reduce GHG emissions. Insulating buildings will not only save energy costs, but also give the building sector an enormous boost and create employment. While some sectors might suffer increased cost, many will seize the opportunity to innovate and get a step ahead of their competitors in adapting to changed market conditions.

5.3.1.2 *Impact of Agriculture on Weather and Climate*

The conversion of native vegetation to Agriculture impacts local and regional climates both directly and indirectly.

The direct impacts

- Large scale differences in land cover give rise to land breezes
- New wet lands to grow crops may result in incidence of damaging frost

- Clearing of forests in large scale significantly increase atmosphere CO_2.

The indirect impacts

- Green revolution involving use of mostly new varieties, fertilizers, pesticides and irrigation over the period 1972-92 contributed to global atmospheric release of nitrogen oxides, nitrous oxide and dinitrogen which are largest human induced sources of GHEs

- Nitrous oxide has approximately 296 times the radiative forcing of CO_2

- The N_2O and CH_4 are contributed to the atmosphere through enteric fermentation by ruminants and manure management

- Over grazing and deforestation contribute to low rainfall

- Exclusive irrigation tends to increased cloudiness thereby increased pests and diseases.

5.3.1.3 *Environmentally effective Policies, Measures and Instruments*

- Financial incentives and regulation for
  - Improved land management
  - Maintaining soil carbon content
  - Efficient use of fertilizers, pesticides and irrigation
  - Increase forest area, reduce deforestation and maintain and manage forests.

The above measures may synergy with sustainable development and with reducing vulnerability to climate change, thereby overcoming barriers to implementation.

5.4   Advises for Major Crops under Climate Change and Variability Scenarios under Asian Conditions

5.4.1   Rice (*Oryza Sativa*)

- Select suitable variety having resistance to at least one major biotic/ abiotic stress for the specific situation and treat the seed with carbendazim @ 4g/kg seed

- Select an area of nursery, which has good irrigation and drainage facility

- Protect the nursery with granules (carbofuran @ 1 kg/5 cents or monocrotophos (1.6 ml/litre) or phosphomedon (1.5 ml/litre) application of cartap hydrochloride 4 @ 8 kg/acre

- Ensure optimum plant population (33, 44 and 66 hills/m^2 for long/mid/short duration varieties respectively) at 2-3 cm depth @ 2-3 seedlings/hill

- Ensure weed control

- Adopt balanced application of fertilizers (N:P:K) on the basis of soil test values

- Avoid top dressing of P or P containing complex fertilizers 15 DAT

- Grow green manure to maintain soil fertility and ensure conjunctive use of organic manures with inorganic fertilizer

- Maintain shallow depth of water (1-2 cm) at the time of transplanting

- Correct micronutrient deficiencies (mainly zinc at 50 kg zinc sulphate/ha in the last puddle or spray 0.2 per cnt ZnSO$_4$ solution 2-3 times at 4-5 days interval in standing crop)

- Crop should not face water stress at panicle initiation, flowering and milk stages

- Adopt integrated pest management practices

- Ensure rodent control on community basis

- Direct seeding with dry or sprouted seed.

5.4.2 Maize (*Zea mays*)

- Use recommended high yielding varieties/hybrids/composites

- Ensure optimum plant population of about 66,000/ha

- Well rotten compost or FYM @ 10-12 t/ha shall be incorporated

- Take up weed management up to one month after sowing

- Atrazine 50% WP @ 2 kg/ha in case of light soils and 3 kg/ha in case of heavy soils is to be mixed in 500 L of water and sprayed uniformly within 2-3 days after sowing on moist soil for management of weeds

- Balanced fertilization on the basis of soil testing

- Zinc application in case of its deficiency at 50 kg zinc sulphate/ha

- A total of 400-500 mm of water would be enough for kharif maize, if water losses through different sources are kept to the minimum
- Intercropping of forage maize in grain maize, up to it's knee-high stage
- In sequence crops, maize-groundnut and maize-redgram to be preferred to maize-maize
- Timely plant protection against stemborer and leaf spots
- Zero till drill machine can be used for dibbling the seed. Plant population should be 83,333 plants/ha.

5.4.3   Jowar (*Sorghum vulgare*)

- Use recommended high yielding varieties/hybrids, seed treatment with Thiram/Captan @ 3g/kg of seed
- Timely sowing before first July to avoid pest problems such as shoot fly and earhead bug
- Seed rate 8-10 kg/ha and spacing 45 × 15 cm
- Proper weed management up to one month after sowing
- Balanced fertilizer management and FYM 10 t/ha
- Flat bed sowing followed by ridging at 30 DAS results in moisture conservation and high yield in red soil
- Sowing with onset of monsoon
- Intercropping Jowar and redgram in 2:1 ratio is more remunerative than sole crop and reduces charcoal not incidence in sorghum and wilt in redgram
- Intercropping of Jowar and cowpea in 2:2 ratio (Jowar in paired rows) enhances the total fodder production without affecting grain yield of Jowar
- Need-based plant protection against insect pests such as shoot-fly, stem borer and earhead bug and also diseases such as sugary disease and grain molds.

5.4.4   Bajra (*Pennisetum typhoides*)

- Use recommended high yielding varieties/hybrids, seed treatment with salt water (2%) and Thiram @ 3g/kg of seed
- Open pollinated varieties (composites and synthetics)
- Sowing by ridge and furrow method will be effective to conserve moisture

- Intercropping of bajra and redgram in 2:1 ratio gives higher monetary returns
- Maintain weed free field and pre emergence application of weedicide Atrazine @ 4g/liter within 48 hours of sowing
- Fertilization @ 60 N, 30 $P_2 O_5$ and 20 K_2O Kg/ha with N in two equal splits – one at sowing/planting and another at about 3-4 weeks after sowing/planting
- Need based plant protection against white ants and root grubs in endemic areas and also against ergot and downy mildew
- Plough the field soon after harvest to bury the ergot inoculums.

5.4.5   Ragi (*Elusine Corocana*)

- Use recommended high yielding varieties, seed treatment with Thiram, Captan or Carbendazim @ 3g/kg of seed
- Minor land smoothening, before sowing helps in better *insitu* moisture conservation
- Nursery – use 5.0 kg seed in 400 m^2 area to plant one hectare.
- Weed management in nursery and also up to 3 weeks after planting
- Ferilizer application @ 60 N, 40 P_2O_5 and 30 K_2O kg/ha, with N in two equal splits – one at planting and another 3-4 weeks after planting
- Spraying of carbendazim 0.05% twice – at 50% earhead emergence and also at complete earhead emergence against blast.

5.4.6   Korra (*Setaria italica*)

- Use recommended high yielding varieties
- Seed treatment with carbendazim @ 2 gm/kg seed
- Weed management up to 30 days after sowing.
- Fertilization @ 40 N and 20 P_2O_5 kg/ha with N in two equal splits – one at sowing and another 3-4 weeks after sowing.
- Intercropping korra + redgram in 5:1.
- In irrigated crop, irrigation should be given after sowing, tillering, ear head emergence, flowering and grain filling stages.

5.4.7 Redgram (*Cajarus cajan*)

- The integrated pest management of *helicoverpa* on redgram summer ploughing, seed treatment with rhizobium culture, following crop rotation and growing intercrops, adjustment of sowing dates with varieties of different durations is recommended

- Release of trichogrammea twice at weekly intervals @ 6500/ha also helps in escaping from helicoverpa
- Adopt IPM against pod borer on redgram
- Intercrop redgram with greengram/blackgram/groundnut/sesamum/ soybean/sorghum or maize
- Under delayed monsoon conditions, grow redgram as sole crop using high seed rate
- Rabi crops may be intercropped with rabi redgram.

5.4.8 Blackgram (*Vigna mungo*)

- Do not cultivate blackgram on light soils in uncertain rainfall areas as it is highly sensitive to moisture stress
- Rabi blackgram under ID conditions and rice follows
- Use recommended high yielding varieties, seed treatment with Captan/Thiram/Mancozeb/Carbendazim @ 3g per kg of seed
- Maintain optimum plant population of 30-35 plants/sq.mt
- Spray 1.5-2.0% urea twice at 40 DAS and 50 DAS
- Follow IPM for the management of tobacco caterpillar
- Timely pest and disease mangement
- Spray twice Mancozeb (0.3%) or Copper Oxychloride (0.3%) at 10 days interval to manage foliar fungal diseases
- Spray twice with karathane (0.1%) + Mancozeb (0.3%) or tridemorph twice at weekly intervals at 50-55 DAS to control the rust
- Foliar nutrition of KNO_3 @ 10 g/L in saline soils.

5.4.9 Greengram (*Vigna radiata*)

- Take up greengram as catch crop before rice planting
- Adopt line sowing and maintain 30-35 plants per sq.mt
- Use recommended high yielding varieties, seed treatment with Captan/Thiram/Mancozeb/Carbendazim @ 3 g per kg of seed
- Apply 20 kg N, 50 kg P_2O_5/ha basally and incorporate
- Spray 1.5-2.0% urea twice at 30 DAS and 40 DAS
- Grow YMV resistant varieties.
- Light irrigations are always beneficial.

5.4.10 Bengalgram (*Cicer arietinum*)

- Use recommended high yielding varieties, treat the seed with Captan or Thiram (3 g/kg seed)
- Grow wilt no need of, stunt and dry root rot resistant varieties
- Follow IPM for the management of *Heliothis*
- Intercultivate twice at 20 and 30 DAS
- Spray Fluchloralin at 2.5 I/ha as pre-sowing incorporated or spray Pendimethalin at 3.3 to 5 I/ha immediately after the sowing or next day.

5.4.11 Soybean (*Glycine max*)

- Use recommended high yielding varieties
- Treat the seed with *Rhizobium japonicum*
- Grow soybean as an intercrop in cotton. This not only increases the area under soybean but also helps in the buildup of predators of Helicoverpa
- Cultivate soybean in rabi under ID conditions.

5.4.12 Groundnut (*Arachis hypogaea*)

- Intercrop with redgram/castor/bajra/jowar in 7:1 ratio
- Deep summer ploughing
- Sow boarder crop (4 rows) of jowar or bajra
- Adopt quality seed of HYV
- Adopt recommended improved variety, plant population ($33/m^2$) and plant protection
- Seed treatment and use small seed without shrivelling
- Do not grow sunflower and marigold which are highly susceptible to tobacco streak virus that causes peanut stem necrosis.
- Use graded seed, follow seed treatment and adopt recommended seed rate.
- Use gypsum (dose 500 kg/ha in the next ploughing or at 45 DAS) and SSP to provide calclium and sulphur

- Avoid delayed groundnut sowings beyond second fortnight of July which will result in increased pest and disease problems and reduced yields
- Practice crop rotation and inter cropping
- Use mechanization for sowing, inter cultivation, harvesting and strippling to reduce cost of cultivation.

5.4.13 Castor (*Recinus cammunis*)

- Use recommended variety/hybrid
- Follow seed treatment (3 gm/kg of seed with Thiram or captan)
- Apply recommended fertilizers, N in 2-3 splits
- Control RHC semilooper, spodoptera through IPM
- Intercrop with redgram and cowpea
- Remove and destroy botrytis affected spikes and spray 0.1% carbendazim solution on the remaining spikes. Apply 20 kg N and 10 kg K_2O/ha if there is moisture in soil
- Popularize rabi castor in non-traditional areas.

5.4.14 Sesamum (*Sesamum indicum*)

- Use quality seed of recommended varieties, seed treatment is essential
- Sow timely in rows, 2 kg seed mixed with 6 kg sand/ac with seed drill, after 2-4 ploughngs and leveling with 2 harrowings, adopting 30×15 cm spacing
- Complete thinning and weeding by 15-20 days after sowing
- Apply fertilizers in split doses. Basal application of 16 : 24 : 8 kg NPK/ac and 20-25 kg urea at 30 DAS
- Go for timely control of leaf folder and gall fly
- Spray wettable sulphur @ 2.0 g/litre of water to control powdery mildew
- Destroy phyllody affected plants.

5.4.15 Sunflower (*Helianthus annus*)

- Use quality seed of recommended varieties/hybrids, seed treatment with 2-3 grams of thiram/captan per kg of seed
- Apply recommended fertilizers; N in 2-3 splits

- Don't use any insecticide during flower opening
- Protect crop from *Helicoverpa*, Spodoptera and parrots. To control spodoptera spray neem oil in early stage @ 5 ml/Lt or monocrotophos 2.0 ml/Lt.

5.4.16 Cotton (*Gossypium sps*)

- Resort to summer ploughing. Avoid mono cropping of cotton
- Use delinted and treated seed of recommended varieties/hybrids
- Go for intercropping with soybean/greengram/cowpea to reduce pest load/improve soil fertility and net returns
- Adopt stem application of systemic insecticides at early stage to encourage defender population
- Collect and destroy eggs of helicoverpa and spodoptera and grown up caterpillars
- Follow IPM schedule
- Remove cotton stubbles soon after final picking
- Do not grow cotton in light soils and chalka and dubba soils as a rainfed or even as an irrigated crop
- Seed cotton from fully opened bolls should be collected during cooler times of the day
- Seed cotton damaged by bollworms should be picked separately
- Seed cotton should be graded and stored in heaps or in gunny bags in dry and well ventilated godowns
- Watering the seed cotton before weighment should be avoided
- Mixing of seed cotton of different varieties should be avoided
- Proper packing should be done to protect from contamination, dampness and fiber quality.

5.4.17 Mesta (*Corchorus sps*)

- Control foot and stem rot disease by seed treatment with Mancozeb at 3 g/kg of seed and soil drenching with 0.2% solution of the same chemical
- Spraying urea solution of 1.25% concentration a day before keeping the sticks for retting for reducing retting period by 12 days in December
- Go for value added products for better returns.

5.4.18 Sugarcane (*Saccharum officianarum*)

- Use short crop for seed material. Soaking setts in 10% lime solution for 60 minutes is beneficial for better seedling establishment and growth under limited irrigation sources
- Give hot water treatment (52^0C for 30 minutes) followed by chemical treatment of seed material (0.1% malathion + 0.05% carbendazim)
- Gap filling has to be done within 2 weeks of ratooning
- Apply P as basal and N and K fertilizers in two splits in recommended doses
- Irrigate at weekly intervals at formative phase during summer
- Ratoon crop management
  - Take up stubble shaving, gap filling, interculture, higher dose of nitrogen, foliar spraying of $FeSO_4$
  - Apply potassic fertilizers to rainfed crop (120 kg./ha) at planting and at 90 DAP
  - Adopt trash mulching @ 3 t/ha at 3rd day after planting, or immediately after rationing to conserve moisture
  - Irrigate once at 10 DAP (life saving irrigation) and second at 30 days thereafter under rainfed conditions.

Early Shoot Borer

- Adopt trash mulching 3 t/ha at 3rd day after planting
- Apply phorate 10 G @ 15 kg/ha at the time of planting
- Irrigate at closer intervals during summer
- Spray chloropyriphos 20Ec @ 2.5 ml/Lt of water at 4th, 6th, 9th weeks after planting.

Scale insect

- Dipping setts in 0.1% malathion or dimethoate (0.05%) before planting
- Detrash keeping eight green leaves at the top followed by Malthion (0.1%), Chlorpyrifos (0.05%) or Dimethoate (0.05%) in first week of July, August and September.

Red Rot

- Use red rot resistant varieties.

Smut

- Give hot water treatment of seed material
- Avoid second ratoon, from the smut affected plant crop.

5.4.19 Chillies (*Capsicum fruitisense*)

- Grow only virus resistant varieties
- Treat the seed with Mancozeb or Captan or Bavistin @ 3 g/kg of seed
- Drench with Bordeaux mixture (1%) or Copper oxychloride (3 g/l) on 13th and 20th day of sowing to prevent damping off diseases
- Follow Integrated Nutrient Management
- Avoid indiscriminate use of pesticides to save the crop from phytotoxicity and residues on the produce
- Avoid repeated use of synthetic pyrethroids as it leads to secondary infestation of sucking pests
- Spray 1% urea 3-4 times at fortnightly interval under moisture stress condition.

6

Weather Health – Crops – Farmers

When born, we have not brought anything with us. We will not take anything along with us when we go back. What we are using today belonged to someone else yesterday and it belongs to someone else tomorrow. Change is inevitable. Accept it. Do not be greedy. We have no right to spoil anything available on the earth.

-Bhavadgita

6.1 Weather Health

The association among "Soil" "Plant" and "Weather (atmosphere)" by virtue of a common process or component in agricultural crop production is known as "Soil-plant- weather continuum". For agricultural purposes, "Soil" is defined as "A dynamic natural body or the solid portion on the surface of the earth in which plants grow". Agricultural crop is defined as "Plants carefully selected and developed over many years and sown on cultivable land to produce food for humans and animals". The term "Weather" is defined as "A state or condition of the atmosphere at a particular place and given instant of time". Crop production strategies involve all these three components.

6.1.1 Murthy's "Weather Health" Concept

Dr. Murthy, the author of the book, introduced and defined the term "Weather health", for the first time in the world".

In "Soil-Plant- Weather continuum" both soil and plant contain "Water" and "Air". It was established that these two have "Life" and scientific terms "Soil health" and "Plant health" are in vogue. However, both "Water" and "Air" are "Weather elements". Therefore, weather has "Life" and "Weather health" for crop production is defined as "The

174

potential force through which weather elements perform their several and cooperative functions optimally for better crop health to produce potential yields". It was observed that to further determine that weather has "Life" there by "Health" all the weather elements have

- Characteristic state or condition
- Constitute existence
- Participate and facilitate metabolism, growth and reproduction
- Responsive to stimuli.

The hypothesis: Weather has "Life" there by "Health". If the weather health is good/ optimum, then optimum agricultural crop yields are possible and vice- versa.

6.1.2 Murthy's "Daily Weather and Agriculture" Concept

To observe the "Weather health" the "Daily weather and agriculture" concept can successfully be used which is both an agricultural meteorological "Tool" and "Service". In this concept the farmers observe "Weather health" as follows:

- They collect daily weather data available in the news papers along with pictorial diagrams and paste on white sheets chronologically

- In addition, the farmers also obtain information on weather from radio, television, internet and mobile telephones (where ever available) and record the same at appropriate points on the white sheets and observe its influence on crops

- After observing the trends of weather and its influence on crops the "Weather health" is determined. Based on "Weather health" the management options for all agricultural operations viz., ploughing, fertilizer application, sowing, inter-cultivation, spraying of chemicals and top dressing of fertilizers will be adopted by the farmers

- Enough recommendations are made available in the book entitled "Weather- Agriculture", "Technical handouts" etc., in local language.

This operational agricultural meteorological tool and service involves "No money" because the news papers are bought by the farmers/villagers for learning and enriching themselves on several issues (political, entertainment and medical). In Asia daily news papers are very inexpensive.

6.1.3   Murthy's Comparison Concept

The Comparison Concept takes into account the past 7-10 days of weather/climate as also the forecast for 7-10 days issued, their (past and forecast) derived parameters (GDD/HTU/PTU) as the basis for forewarning. These derived parameters are compared with the scenarios of past seasons or years and a suitable set of common similarities on crop yield, incidence and vigor of pests and diseases and their influence on crop performance are arrived. This scientific information helps to determine both ongoing and future scenarios of occurrence/incidence and vigor of pests and diseases and crop yields. This concept is also useful to develop thumb rules. Farmers adopt appropriate management options based on technical handouts.

6.1.4   Growing Degree Days

In "Weather health" and "Daily weather and agriculture" concepts, as also in the preparation of "Weather- Agriculture" book and technical handouts in local Telugu language, in addition to rainfall probabilities the GDD and HTU were successfully used.

By definition "Degree days are summation of mean temperatures over a base temperature". The GDD is also known as "Heat units" "Thermal units" "Effective heat units" "Growth units" etc. The accumulations are made on daily basis and are also accumulated between any two phenological events of crop plants or dates.

The concept assumes that

- There is a direct and linear relationship between growth of a crop plant and air temperature
- A crop requires a definite amount of accumulated heat energy for optimum crop yields
- The biotic potential of an agricultural crop plant is dependent on the heat requirement for its growth, development, reproduction and grain yield.

6.1.5   The Canonical form for Calculating GDD is

Degree Days $(^{0}D) = \{(T \max + T \min)/2 - T \text{ base}\}$

where

- "T max" and "T min" represent the daily maximum and minimum temperatures respectively

- "T base" is the base temperature.

Usually Degree Days are expressed as "^{0}D" to distinguish from temperature units.

The Base Temperature

The base temperature is one below which the internal metabolism activities of crop plant cease to function. Though the base temperature varies from crop to crop, it is constant for a specific crop. In India during *Kharif* season (South West monsoon crop season) the base temperature is taken as 10 degrees centigrade and for *Rabi* (North east monsoon season crop) the base temperature is taken as 5 degrees centigrade.

Advantages/Importance of GDD

The GDD is a small and simple concept of relating plant growth, development and maturity to the air temperature. The growth of plant is dependent on the total amount of heat to which it is subjected during its life time. The GDD are useful in many ways:

- In guiding all the agricultural operations and land use planning
- To forecast crop harvest dates, yield and quality
- In forecasting labour required for agricultural operations
- Introduction of new genotypes in new areas
- In predicting the likelihood of successful growth of a crop in new areas.

Modification to GDD

To further enhance the biological meaning and wider area coverage of GDD applications in "Weather health" and "Daily weather and agriculture" concepts the following modifications are suggested:

- Converting GDD into HTU and PTU
    - Helio thermal units: GDD × number of actual sunshine hours
    - Photo thermal units: GDD × day length (hours the product of degree day and day length or any day)
- Incorporating an upper temperature threshold
- Using only the maximum or minimum temperature or position of the day.

Incorporating functions for other environmental factors that affect phenology or the process being considered.

6.2   Crops

6.2.1   Rice (*Oryza sativa* (L.))

6.2.1.1 *Preamble*

Botany of Rice

Genetics show that rice was first domesticated in YANGTZE river valley. The earliest remains of cultivated rice in India have been found in north and west since 2000 BC. In Africa, it has been cultivated for 3500 years in Niger river delta and extended to Senegal, Morocco, East Africa and Sub-sharan Africa. In Middle East rice was introduced in *hellenistic* times. In Europe, it was introduced in tenth century and spread to Italy and France after the middle of 15th century. In 1520s rice was introduced to Caribbean South America.   In 1694 rice was introduced in South Carolina and it has been grown in southern Arkansas, Louisiana and Texas since mid 1800s. Although attempts to grow rice in the well watered north of Australia have been made for many years, they have consistently failed because of iron and manganese toxicities and destruction by pests. So, since 1920s it is grown as irrigated crop in Australia.

Rice is the seed of a monocot plant *Oryza sativa* (L.) and belongs to *graminacea* family. As a cereal grain it is the most important staple food especially in east, south and south East Asia, the Middle East, Latin America, Africa and the Westindies.

- It is normally grown as a seasonal (100-180 days duration) or annual plant

- It also survives as a perennial and produce ratoon crop for 30 years

- The rice plant grow to 1-1.8 meters tall depending on the variety and soil fertility

- It has long and slender leaves of 50-60 cm long and 2-2.5 cm width

- The small wind pollinated flowers are produced in a branched arching to pendulous inflorescence of 30-50 cm long

- The edible seed is a grain (caryopsis) 5-12 mm long and 2-3 mm thick.

Importance of Rice

Rice cultivation is well suited to countries and regions with high rainfall because it requires plenty of water for cultivation. It can be grown practically anywhere viz., plain land, flooded water and steep hill.

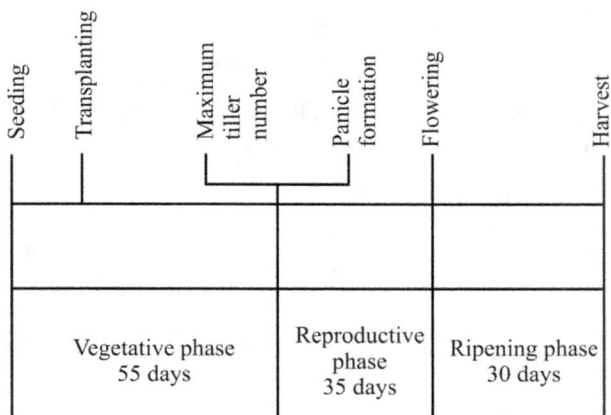

- Days in vegetative phase differ with variety
- Reproductive and ripening phases are constant for most varieties
- Sowing to harvest may be 180 days or more for long duration varieties.

Fig. 6.1 Growth stages of the rice plant of 120 days duration

- The United Nations General Assembly (UNGA) during its 57th session on 16th December 2002 declared 2004, the International Year of Rice (IYR). This dedication of an international year to rice, a single crop, is unprecedented in the UNGA's history

- Rice is not only a fundamental commodity but also a primary food source, considered as "Global food"

- It influences the issues of global concern such as food security, poverty alleviation, preservation of cultural heritage and sustainable rural development

- It is the staple food for over half of the world's population

- In Asia alone, more than 2 billion people obtain 60-70 percent of their calorie intake from rice and its derived food products

- More than one fifth of calories consumed worldwide in human nutrition is provided by rice

- Rice is a rapidly growing food source in Africa

- Almost a billion households in Asia and America depend on rice systems for their main source of employment and livelihood

- About four fifths of the world's rice is produced by small scale farmers and is consumed locally

- Rice systems support a wide variety of plants and animals, which also help supplement rural diets and incomes
- Rice is on the frontline in fight against world hunger and poverty
- Rice is a symbol of both cultural identity and global unity
- For all the reasons "RICE IS LIFE".

Uses of Rice

Rice is a staple food. It is used by many ways as detailed below

- Cooking of rice is the most popular way of eating
- Rice starch is used in making icecream, puddings, gel and distillation of potable alcohol
- Rice bran is used in confectionary products like bread, snacks, cookies and biscuits
- Rice bran oil is used as edible oil and in manufacturing of soap and fatty acids
- Flaked rice and parched rice are made from parboiled rice and used for eating
- Puffed rice is made from paddy and used for eating
- Rice husk is used as a fuel, building and packing material
- Broken rice is used in making rice floor, noodles, poultry feed and rice cakes
- Rice straw is used as animal feed, fuel, mushroom bed, mulching, paper manufacturing and compost preparation
- Paddy is used as seed.

Global Rice area and Production Scenario

- Rice is grown worldwide over an area of 155 million hectares with an annual production of 673 million tonnes
- It is grown in 176 countries in the world
- Among all the crops it is highest in global production but second to wheat (214 million ha) in global area
- During the last 50 years, the global area of rice has increased more than one and half time and the production has increased four times while the productivity exceeded two times

- Continent wise, more than 90% of the rice is produced and consumed in Asia (It is biologically and culturally bound with the lives of Asians)
- The other two continents growing and consuming more rice are Africa and Latin America
- In North America, Europe and Australia, rice is grown in a very limited area, though the productivity is high
- Among the predominant rice growing countries, China is the world's biggest producer of rice (185 m.t.) followed by India (130 m.t)
- The other major rice growing countries include Indonesia, Bangladesh, Vietnam, Thailand, Myanmar, Philippines, Japan, etc.
- More than 90% of rice is produced and consumed in Asian countries
- Even though china is second in area next to India, it occupies the first place (6.33 t/ha) ahead of India (3.0 t/ha) in production
- One of the main reasons for higher production is favourable weather and climate of China in addition to rich soils and high percentage of irrigated area
- The major reasons for low productivity in India are the losses due to cloudy weather in rainy season, insect pests and diseases caused by weather changes and weeds.

6. 2.1.2 *Rice Production Ecosystems*

Rice culture is divided into two broad groups viz., upland and lowland rice cultures. However, globally there are many "Rice production ecosystems" that fall under these groups. Technological recommendations for growing rice under these cultures, System of Rice Intensification (SRI) and arobic rice are described below.

Major Global Rice Ecosystems

Lowland Rice Ecosystem

The term "Lowland rice" refers to growing rice on flat land under controlled supply of water (Irrigation). Lowland rice is also known as "Irrigated rice" "Water logged rice" "Flooded rice" etc. The environmental conditions are stable and uniform. The pest and disease incidence is low and weeds are not serious problem. Even though both the input requirement and cost of production is high, the yields are not only high but also stable.

- The crop is either transplanted or direct sown
- Variability in the onset of monsoon is a factor that determines the beginning of planting because water is required for land preparation and soil puddling
- Variability of rainfall affects the rice crop at different times
- If the variability is associated with the onset of rain, the stand establishment and the growth duration of rice are affected
- If variability is associated with cessation of monsoon at the reproductive or ripening stage of the rice crop, yield reduction is severe
- The World's two largest rice growing countries viz., India and China have many areas that receive around 1350 mm average rainfall
- India often has inadequate or excess rainfall during the rainy season. As a result drought or flood and sometimes both cause substantial damage to rice production
- Due to specific weather, climate and agronomic conditions in most soil types rice is sown in lowlands by broadcasting seed in dry or moist fields before the onset of monsoon
- Land preparation for direct seeding starts with onset of monsoon
- To take up direct seeding of rice a few rain free days after puddling are needed. Otherwise the seeded rice gets washed away or collected near bunds which results in poorest plant stand
- 90-100 kg/ha of pre germinated seeds are broadcast
- Application of NPK fertilizers (@80-45-30 kg/ha), proper weed control, need based application of pesticides are essential for optimum grain and straw yields.

Upland Rice Ecosystem

The term "Upland rice" refers to growing rice depending on rainfall for moisture and sloping fields which are prepared and seeded under dryland conditions. "Upland rice" is also known as "Rainfed rice". Brazil is the largest producer of upland rice in the world. The environmental conditions are unstable and variable. The pest and disease incidence is high and weeds are a serious problem. The cultivation is also characterized by low input cost and low cost of production. Compared to lowland rice ecosystem the yields are low and unstable because of relatively low tillering and water supply is through rainfall alone.

- Rainfall variability is further more critical for upland rice than lowland rice
- Moisture stress can damage rice crop yield up to 30% if 200 mm of precipitation occur in 2 days and then if no rain occurs for the next 20 days
- In contrast, an evenly distributed rainfall of 100 mm/month is preferable to 200 mm/month that falls in two days
- If the rainfall pattern of upland rice growing areas, is 200 mm/month as base, then:
    - It is recommended that varieties that mature in less than 100 days may be preferred for unimodel rainfall pattern.
      Example: Uiberaba (Brazil)
    - If the rainfall pattern is also unimodel, but, duration is longer, then varieties that mature in 100–150 days perform well (*Example*: Bukidnon in Philippines)
    - Another option in areas like Bukidnon is planting a short duration upland rice followed by another short duration cash crop (onion, mungbean).
- In Braizil droughts of 5-20 days duration occur in Cerrado region. The high frequency of drought periods during the crop season is known as "Veronica". Upland rice in Cerrado region is adversely affected particularly at reproductive and ripening stages
- Upland rice is grown in areas of high rainfall (Assam and West Bengal in India, most of Bangladesh) and also in low rainfall areas (Madhya Pradesh in India)
- In upland areas of Myanmar rainfall is as low as 500 mm and as high as 2000 mm
- In west Africa, where rice is grown mostly as upland crop, the amount and distribution of rainfall are of paramount importance
- The rain may begin any time from March to July (It begins later in high latitudes)
- The rainfall is unimodel (having one peak) in areas with short rainy season. It is bimodal (having two peaks) with 1 to 2 month break from July to August in areas with long rainy season
- The regions of bimodel rainfall includes southeastern Ivory coast, southern Ghana, Southern Togo and Benin and southern Nigeria

- Less important areas of bimodel distribution include southeastern Guinea and northeastern Liberia

- In west African areas of less than 1000 mm annual rainfall, the season may begin as early as late March

- For Latin America, a 1000 mm rainfall with 200 mm of monthly rainfall during the growing season is adequate for growing upland rice

- Brazil, which by far the largest upland rice growing country in Latin America, has a distinct rainy season that begins in October and ends in April

- The annual rainfall varies from 1300 to 1800 mm and 70-80% of the rain falls during the upland rice growing season

- The rainfall becomes less in February, leading to an ideal harvesting time

- Rainfall in most of Peru's Amazon Basin ranges from 2000 to 4000 mm annually

- In many Central American countries more than 2000 mm of annual rainfall is common. This amount of rainfall is more than enough to grow one upland rice crop.

Irrigated Rice Ecosystem

Water is made available from different sources of irrigation viz., canals, wells and tanks. Soil is puddled to form impermeable layer that helps in reduced percolation of irrigation water.

- In most areas of assured irrigation, transplanting is done using seedlings raised in nurseries

- Direct seeding of pre germinated rice is followed in limited areas

- Paddy row seeder is recommended for effective direct seeding in puddled soils. The methods followed for raising the nurseries are wet nursery, dry nursery and mat (depog) nursery

- 20 grams of well filled dense seeds are broadcast for every square meter

- During winter season double the dose of phosphorus fertilizer is necessary

- Field is dry ploughed three weeks before planting time and submerged by inundating with 5-10 cm of standing water

- Application of 10 tonnes of farm yard manure is recommended before planting time and after its proper decomposition, the main field has to be levelled after applying basal dose of N, P and K fertilizers

- 22-25 days old seedlings (4-5 leaf stage) are good for transplanting at shallow depth of 3-4 cm

- 2-5 cm water shall be maintained throughout the growing season

- Draining out of water 15-20 days after 50% flowering is recommended to ensure fast ripening of seeds and for ease in mechanical harvest

- In most of the temperate rice growing countries in Asia (Japan, Korea and China) and other regions such as North America, Australia and Europe rice growing is determined primarily by temperature pattern. Rice is grown under irrigated ecosystem in these areas

- With irrigation, planting can be adjusted to take advantage of optimum temperature and high solar radiation.

Hill Rice Ecosystem

- Considerable areas in the world are under hill rice ecosystem (up to 2300 meters on above MSL)

- Rice is either transplanted or direct seeded. The production constrains are "Low temperature" and "Drought" right from seeding. Low temperatures results in high spikelet sterility during reproductive phase, poor panicle exertion and short growth duration

- Application of nitrogen is not recommended if very cool weather prevails at panicle initiation stage.

System of Rice Intensification (SRI)

- SRI refers to a particular set of innovative practices which improve rice plant health and yield

- SRI is presently followed in more than 20 countries as a method of rice cultivation which has potential to produce more rice with less water. It is projected as a revolutionary technology in a sense that it tries to change traditional practices especially with respect to water management that existed for thousands of years

- The SRI was first developed in Medagascar during 1980's and less known till 1997

- The greatest potential of SRI is seen in its six basic principles
  - Transplanting young (8-12 day old) seedlings singly using 5 kg / ha of seed
  - Careful transplanting at shallow depth
  - Adoption of 25 × 25 cm spacing (square pattern)
  - Water management to keep soil moist by alternate flooding and wetting
  - Use of rotary weeder for interrow weeding
  - Use of organics such as FYM/ Compost / green manure.

Aerobic Rice

- Prepare the land on receipt of rainfall like that of any rainfed crop
- Sow the paddy seed by gorru or behind the plough followed by basal dose of fertilizer application and irrigate
- Irrigate crop at 5-6 days interval
- Pre emergence application of pendimethalin @ 5 ml/l followed by 2 hand weedings
- Weeds major problem
- More top dresses of fertilizer application
- Save water by 400-500 mm.

6.2.1.3   *Influence of Weather Elements on Rice Crop Production*

Among various countries that grow rice, in 3 countries the yields are 6 t/ha; 17 countries 4 t/ha, 78 countries < 3 t/ha; 58 countries < 2 t/ha and in 20 countries it is even < 1 t/ha. This variation is mostly due to the enormous influence of weather elements. Weather elements influence physiological processes and incidence of pests and diseases. Also, differences in rice productivity are largely accounted for the differences in influence of weather elements like solar radiation, temperature, wind, rainfall and humidity.

Effect of Solar Radiation

- Rice is a short day plant and sensitive to photoperiod. Therefore, long days can prevent or considerably delay its flowering (However, rice varieties exhibit a wide range of variation in their degree of sensitivity to photoperiod)
- The Photosynthetically Active Radiation (PAR) lies between 0.4 and 0.7 microns in electromagnetic spectrum. It is also referred as

"Light". The PAR available at the canopy, its interception and utilization by the crop determine to a large extent the yield and quality in rice

- At least 8 hours of sunshine is required for better seedling growth. Such seedlings attain four leaf stage and ready for transplantation in 20-22 days during rainy season and in 40 days during dry season (winter)

- Lesser than 8 hours of sunshine due to shade by clouds cause elongation of leaf blade and leaf sheath which in turn results in taller and weaker seedlings (undesirable character)

- Nurseries raised under or near tall trees, buildings and tall structures produce less dry matter, because of the less quatity solar radiation due to shadows of these objects both in the morning as well as evening. Such seedlings, which contains less dry matter are prone to pests and diseases

- Rice plants develop longer basal internodes under cloudy weather and high plant density. Plants with longer internodes can not support well developed panicles. Therefore, plants fall flat on ground resulting in huge grain losses

- The solar radiation requirement of rice crop differs from one growth stage to other. Shading at vegetative stage slightly affects the subsequent growth stages thereby the yield. In addition to this, stressed environment results in net decrease in crop dry matter

- The most critical period for solar energy requirements for rice plants is from panicle initiation to 10 days before maturity. Increase in dry matter production is directly correlated to high solar radiation during this period

- Shading at reproductive stage, however, has a pronounced negative effect on spikelet number and during ripening reduces the yield considerably because of reduction in percentage of filled spikelets. An average of 300 cal/cm^2/min during reproductive stage makes yields of 5 t/ha possible

- In the tropical regions the solar radiation is higher in the dry season than in the wet season. Consequently, the dry season yields are higher

- The excessively cloudy weather during wet season is often considered a serious limiting factor for rice production in monsoonal Asia

- Photoperiod sensitive varieties have traditionally been grown in monsoonal Asia.   These varieties provide basic stability in rice production

- In some areas the need to delay harvesting until monsoon season flood water has receded makes it essential to grow varieties with long duration

- Photoperiod sensitive rice varieties enable the farmers in tropics and sub tropics to plant rice at any time of the year without much change in growth duration

- Short duration varieties can be planted in any month in the tropics and will mature in a fixed number of days

- However, it is obvious that insensitivity to day length is essential in one situation and a liability in another

- Long day length and high level of solar energy during ripening period also contribute to high grain yield in temperate rice growing regions of the United States, Southern Australia and parts of Europe

- Rice response to solar radiation is greater at higher levels of nitrogen. Also, high solar radiation after panicle initiation gives higher yields

- Another major utility of solar radiation is the solar drying of paddy seeds and husk. Even though this practice is traditional it is more economical than forced convection driers

- The present tendency to produce improved, early maturing photoperiod insensitive varieties which may fit into multiple cropping systems is the characteristic of progressive agriculture.

Effect of Temperature

- Temperature regime generally influences not only the growth but also the growth pattern. During the growing season, the mean temperature, the temperature sum, range, distribution pattern and diurnal changes or a combination of these are highly correlated with grain yields

- Seedlings grow faster at $25\,^{\circ}C$ to $30\,^{\circ}C$. More than 12 hours of sunshine increase vegetative stage which is an undesirable character

- Areas in lower latitudes have highest temperature at sowing time and a slowly declining temperate until maturity. These declining temperatures enable more dry matter production thereby higher grain yields

- Higher grain yields in temperate countries than in tropical countries have generally been attributed to the lower temperatures during ripening. This is because the ripening is extended due to low temperature, giving more time for grain filling

- Generally, high temperature accelerates and low temperature delays heading.

Effects of Low Temperature

- A variety which is susceptible to low temperature when held at $15\,^{\circ}C$ for 4 days during reduction division stage, the sterility percentage was 51. Whereas a low temperature tolerant variety recorded only 5 % sterility

- Temperature as low as $12\,^{\circ}C$ will not induce sterility if they last for 2 days, but, will induce about 100 per cent sterility if it last for 6 days

- Injury due to low temperatures is a major constraint for rice production in hilly areas, tropics and subtropics

- In temperate regions, cold injury is the main constraint limiting the rice growing area and length of the growing season. The major factors that cause cold injury to rice are (a) cool weather and (b) cold irrigation water

The common types of symptoms caused by low temperature are

- Poor germination
- Slow growth and discolouration of seedlings
- Stunted vegetative growth characterized by reduced height and tillering
- Delayed heading
- Incomplete panicle emergence
- Prolonged flowering period because of irregular heading
- Degeneration of spikelets
- Irregular maturity.

Effects of High Temperature

- High temperature is a critical factor in rice grain production. A high percentage of sterility and empty spikelets in rice crop are noticed in Oasis areas of Egypt

- Several authors reported heading to be stage at which rice plant is more sensitive to high temperature

- It is most common to have maximum daily temperatures from 35-41 $^{\circ}$C or higher in semi arid regions and during hot months in tropical Asia. In these regions a heat susceptible variety may suffer and yield less grains

The high temperature effects at different growth stages are

- At reproductive stage: White spikelets white panicles and spikelet sterility

- During anthesis - Sterility

- At ripening - Reduced grain filling

The occurrence of various phenological events and the biomass production depend on the accumulated heat sums, which is having a strong linear relationship.

Effect of Wind

- A gentle wind during growing period improves grain yield as it increases turbulence in the canopy. The air blown around the plants replenishes the carbondioxide supply to the plant

- Gentle winds from 0.75 to 2.25 centimeters per second help in increasing the photosynthesis

- However, strong winds (cyclonic) at heading stage may cause lodging. They often desiccate the panicles, increase both the floret sterility and the number of abortive endosperms

- Strong winds enhance spread of bacterial diseases

- Dry winds also cause desiccation of leaves and mechanical damage to the plants.

Effect of Rainfall

The effects of water shortage and excess water on rice yields are similar to any other crop depending upon the phonological stage of the crop. The term rainfed applies to both upload and lowland rice culture. The effect of rainfall on rice crop are

- The effects include leaf rolling, leaf scorching, impaired tillering, stunting, delayed flowering, spikelet sterility and incomplete grain filling

- Rice crop is most sensitive to water deficit from reduction division to heading stages. A persist of 11 days of drought, 3 days before heading reduced the yields significantly causing 59 to 62 per cent sterility

- Heavy rain causes submergence and flooding of nurseries. As a consequence due to lack of air, root growth is poor. When such seedlings are pulled and transplanted the survival is 60%

- Closer spacing (10 × 8, 10 × 10, 10 × 12 cm) and cloudy days (shadows) cause long leaf sheaths. This is an undesirable plant character. Rice plants with long leaf sheaths are prone to attack and damage by pests and diseases

- In rainy season 20-25% more tillers are produced than in dry season. So, to enhance tiller production, more Nitrogenous (50% more fertilizer) has to be applied in dry season than that recommended in rainy season

- When leaves are wet due to rain or dewfall the applied fertilizer cause leaf burn, as it sticks on to wet surface. So, do not top dress fertilizers when leaves are wet

- The nitrogen use efficiency is more in dry season than in rainy season, because, great losses occur in rainy season due to volatilization and denitrification.

- Rain days (cloudy days) and low temperatures ($<20\ ^{\circ}C$) delay ripening stage while cloudless days (more than 8 - 10 hours of sunshine) and 32-35 °C temperature shorten it

- As the danger of shading due to cloudiness is less in dry season plants do not compete for light and do not grow taller than optimum height. Therefore, lodging is less in dry season than in rainy season

- Tall varieties are not recommended during rainy season because heavy rains and stronger winds (cyclone) lodge the crop

- The grain yield obtained during wet season is lower than that in dry season because of lower levels of solar radiation received during grain filling and ripening stages

- Under rainfed rice cultivation, where temperatures are within the critical ranges. rainfall is the most critical factor limiting rice cultivation

- Rainfed rice cultivation is limited to areas where annual rainfall is around 1350 millimeters

- The variation in amount and distribution of rainfall is the most important factor limiting yields of rainfed rice which constitute about 80 per cent of rice grown in south and south-east Asia

- The amount and distribution pattern of rainfall varies widely from location to location and year to year.

Example: The Deccan Plateau of Indian Sub-continent receives a rainfall of 500 millimeters, which is much less than the water requirement of rice crop (1240 millimeters). Hence, rice is grown when irrigation water is available. But, in Philippines where 2468 mm of rainfall is received per year, from total monsoon (SW and NE monsoons), rainfall is fairly and evenly distributed throughout the year. There is no dry season. Hence, rice is grown throughout the year, but, harvesting and drying are the problems. Whereas, Cuttack in India, receives 1545 mm per year from South-West monsoon only. Rainfall is distributed from June to November evenly and rice crop is grown favourably.

When grains are mature and if it rains the non dormant seeds germinate. So, rice varieties which are dormant are recommended for rainy season.

Effect of Relative Humidity

- Humidity influences the incidence of rice crop insect pests and diseases and hence grain yield i.e., usually more the R.H. more the chances for occurrence of pests and diseases

- Long periods of dew or fog also cause increased incidence of pests and diseases

- High humidity (80-90%) have negative influence on physiological processes thereby result in lesser rice plant growth, development and grain formation

- Differences in rice productivity to a greater extent are also accounted for trends in humidity (30-45% is always good, but, 80-90% are always detrimental).

6.2.1.4   *Influence of Weather on occurrence of Rice Pests and Diseases*

Weather and Rice Pests

Rice crop is prone to several abiotic and biotic stresses among which onslaught from pests and diseases cause yield losses up to 28%. The influence of weather on occurrence of major insect pests that cause damage in most rice growing countries are described below.

Table 6.1 Influence of weather on rice pest outbreak

S.No	Pest	Weather Conditions
1	Stem Borer (SB)	Most abundant in aquatic habitats where flooding occurs and places where multiple rice crops are grown annually. Towards end of rainy season these are seen abundantly and cause economic losses. Adults are quiet during day and are active at night. More damage occurs during drought years in the tropics. Low night temperatures with more than 7 hours of sunshine during day coupled with less than 8 mm/day rainfall increase severity.
2	Gall Midge (GM)	Becomes highly abundant during the rainy season in irrigated and wetland environment of more than 82% RH. Also occur and damage in relatively low level during dry season in irrigated areas when fields are continuously flooded. Abundance is favoured by cloudy or rainy weather (more than 200 mm in a month) and cause extensive drainage.
3	Brown Plant Hopper (BPH)	Mainly a pest of irrigated and wetland rice environments. Hopper burn occurs more rapidly during cloudy weather. High rainfall in SW monsoon (August-September); More than 30 mm rain for at least 2 weeks; warm and humid weather coupled with 21-23 $^{\circ}$C night temperatures for a week increase severity, necessitating emergency control measures.
4	Leaf Folder (LF)	Occurs in all rice growing environments and are more abundant in rainy season. Higher infestation occurs in areas where cloudy weather occurs for 5-7 days and also when the rice crop is shaded by trees. Prolonged dry spell after heavy rains coupled with cloudy weather for at least one week increase severity.
5	Rice Hispa (RH)	Prevalent in wetland rice environments, particularly irrigated areas where rice grows throughout the year. It is more abundant in the rainy season. During the dry season adult numbers in rainfed areas decline.

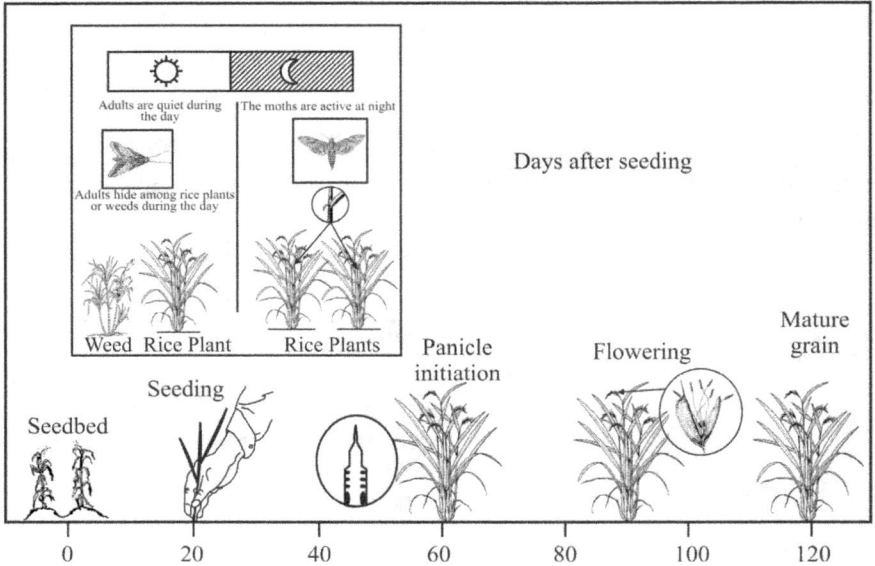

Fig. 6.2 Rice stem borer

Symptoms of Damage of Major Rice Pests

- **Stem Borer (*Scirpophaga incertulas*)**

  Among five species of stem borers yellow stem borer (YSB) is the most wide spread, dominant and destructive pest. It causes damage throughout the growth period. Affected plants show characteristic symptoms of dead heart (white ears). Infestation results in partial chaffiness of the glumes and ill filled grains. Larve cause broad longitudinal whitish dis-colouration on leaf surface (Fig.6.2).

- **Gall Midge (*Orseolia oryzae*)**

  This is a key pest that causes damage in all ecosystems of rice cultivation. The adult gall midge is a mosquito like insect. The characteristic damage is elongation of leaf sheath called "Gall". The gall which resembles an onion leaf glistens in the field, hence is called "Silver shoot". An unwanted profuse tillering and stunting of plants are associated with gall formation (Fig.6.3).

- **Brown Plant Hopper (BPH) (*Nilaparvata lugens*)**

  Both nymphs and adults of this pest cause direct damage by sucking plant sap leading to stunted growth and reduced tillering. In case of severe infestation the field gives a burnt appearance known as "Hopper burn". Apart from direct damage the BPH is also a vector of grassy stunt virus (Fig.6.4).

- **Leaf Folder (*Cnaphalocrocis medinalis*)**

  The larve of this pest fold the leaves longitudinally and feed within, resulting in a linear pale white stripe. In cases of severe infestation, the leaf margins and tips are totally dried up and the crop gives a whitish appearance (Fig.6.5).

- **Rice Hispa (*Dicladispa armigera Oliver*)**

  Mainly during the vegetative stage both adults and grubs feed on leaves. In severe epidemics the leaves dry up and crop looks like a scorched appearance (Fig.6.6).

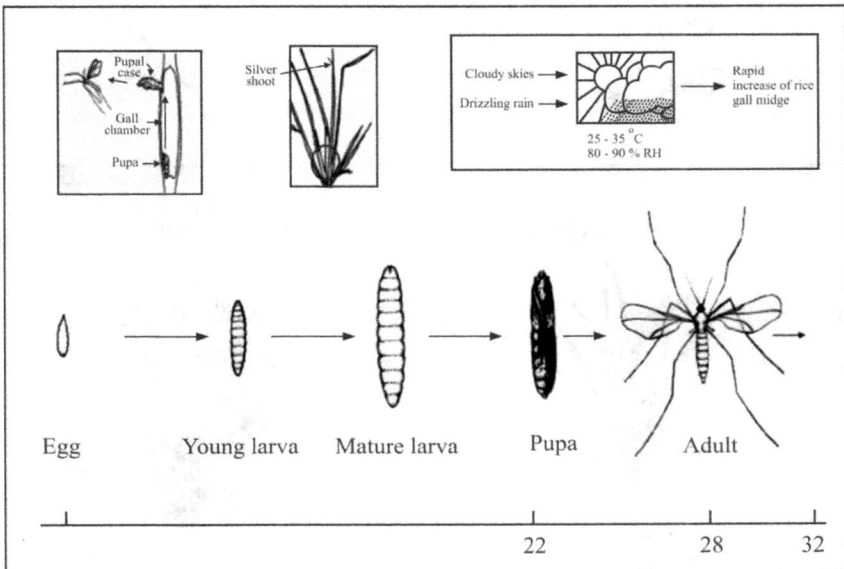

Fig. 6.3 Development of rice gall midge

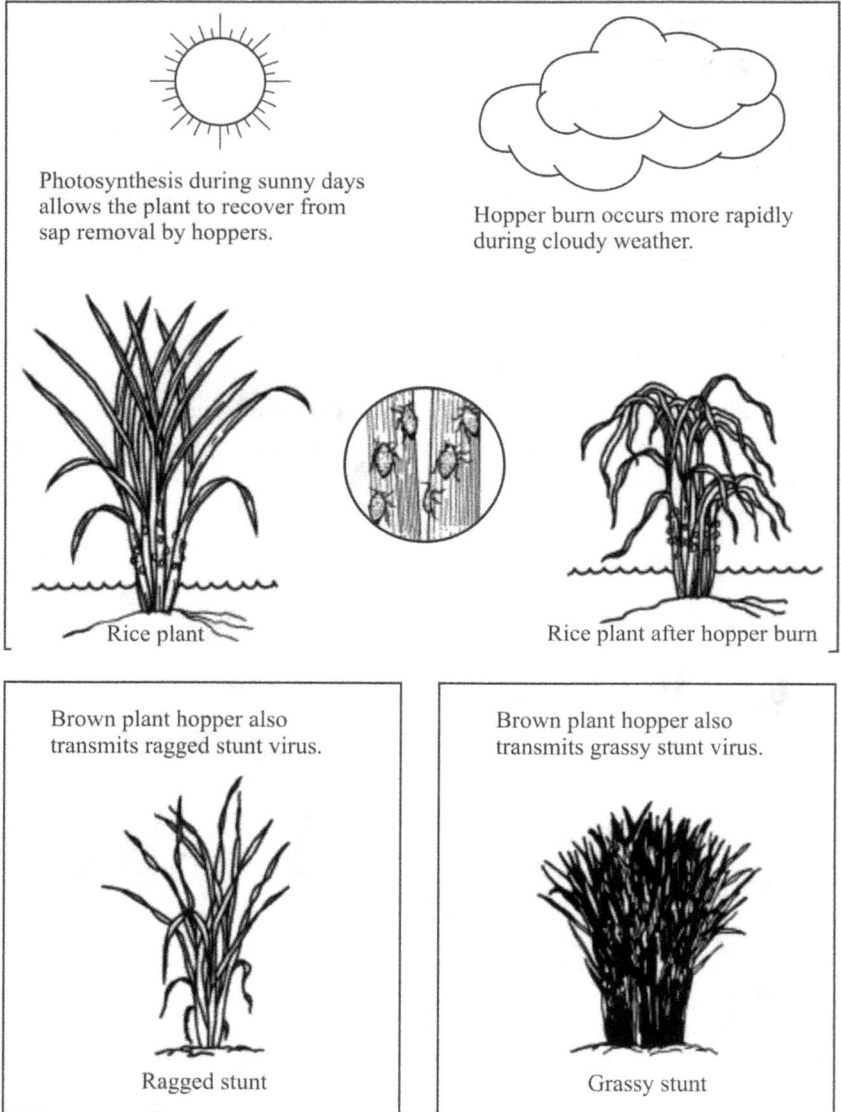

Photosynthesis during sunny days allows the plant to recover from sap removal by hoppers.

Hopper burn occurs more rapidly during cloudy weather.

Rice plant

Rice plant after hopper burn

Brown plant hopper also transmits ragged stunt virus.

Brown plant hopper also transmits grassy stunt virus.

Ragged stunt

Grassy stunt

Fig. 6.4 Development of rice brown plant hopper

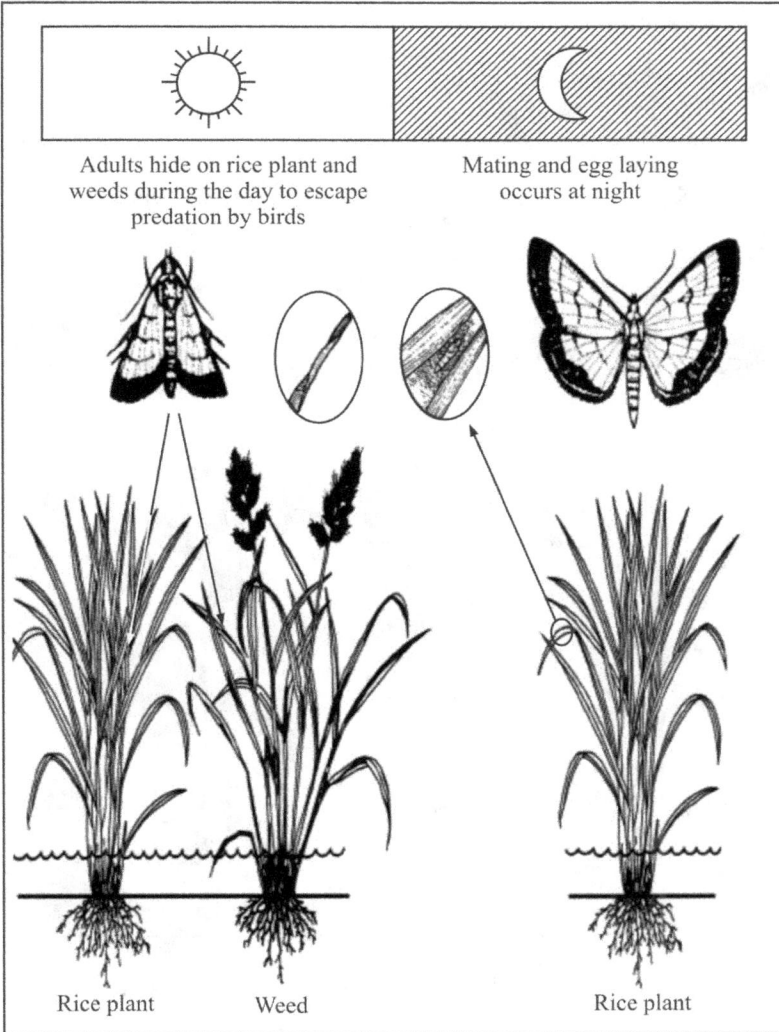

Fig. 6.5 Rice leaf folder

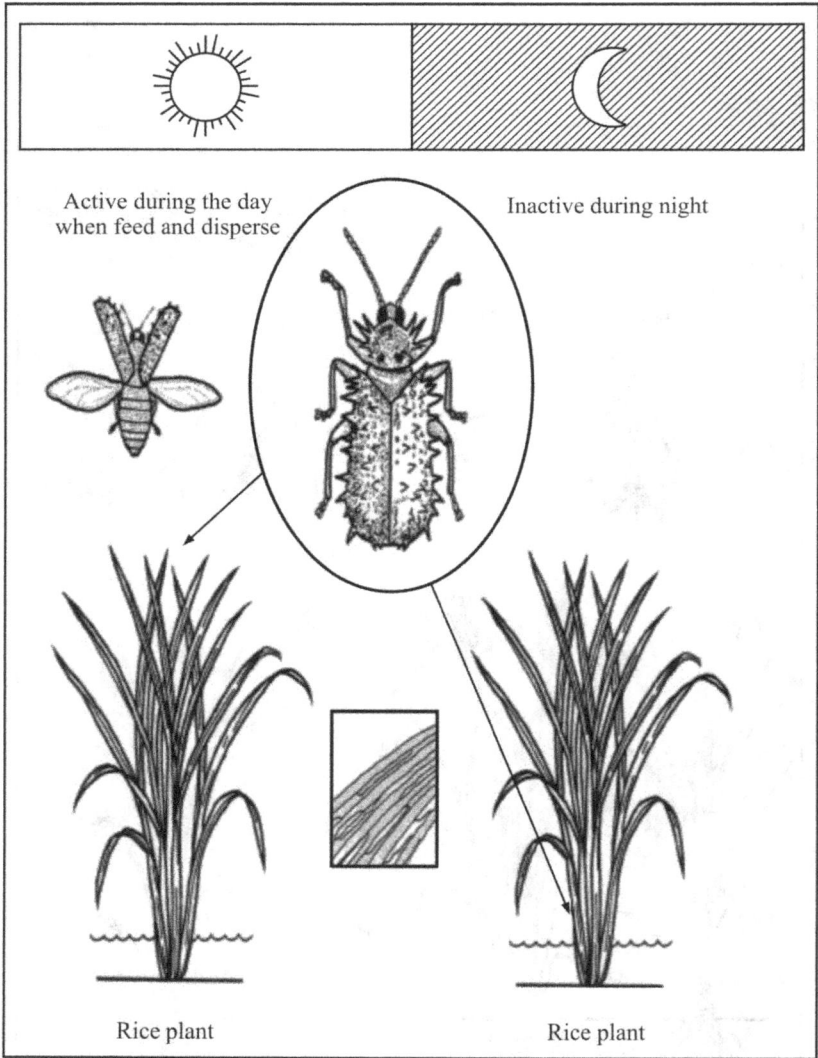

Fig. 6.6 Rice hispa

Integrated Rice Pest Management (IPM) (I – Increases yields P- Protects Environment M- Manages pests)

To obtain sustainable rice crop yields with least damage to environment the IPM is the most appropriate approach. Also, to reduce toxic hazards of pesticides, IPM is useful. Pests developing resistance to pesticides can successfully be minimized or eliminated.   For all the countries the following information is applicable.

- Pest surveillance (monitoring)
- Install light and pheromone traps @10/ha

Cultural Practices

- Do not grow dense nurseries which lead to weak growth
- Yellow stem borer's spread can be reduced by summer ploughing
- Formation of alley ways of 20 cm for every 2 meters reduce onslaught of pests and diseases in nurseries
- Brown plant hopper damage can be reduced by moderate doses of nitrogenous fertilizers as top dressing
- Seedling root dip operations shall be done in the evening
- Timely transplanting reduces incidence of yellow stem borer and gall midge
- Maintain 3-5 cm of water depth in the main field
- Avoid late planting
- Grow resistant varieties.

Mechanical Control

- Collect and destruct as many as possible of egg masses, larvae and infested plant parts
- Clip the tips of seedlings before transplanting to remove the egg masses of yellow stem borers.

Biological Control

- Release Trichogramma @ 1,00,000 per hectare
- Provide proper environment for multiplication of natural enemies.

Chemical Control

- When essential spray monoprotophos @ 1.6 ml/l
- Avoid sprays during noon time
- Avoid carbofuron in the main field
- In locations of low temperatures (15-25 °C) apply double the dose of phosphorous in 2-3 dressings

- Protect against bird damage of seed by netting or tying colour ribbons
- Plough land during summer to reduce weed growth and to store more rain water
- Ensure timely planting of rice variety specific to location, season and availability of resources
- Drain out water before top dressing with nitrogenous fertilizers and let in water after 24 hours
- Drain water one week prior to harvest/maturity
- Repeat spray, if rain occurs within 3 hours after spraying
- Planting early at the beginning of the monsoon rains is a method that allows a field to escape hispa build up on alternate hosts or other rice fields.

Weather and Rice Diseases

The rice crop suffers from three major groups of diseases namely fungal, bacterial and viral. Of them the influence of weather on most important 3 fungal (blast, sheathe blight, brown leaf spot) and 2 bacterial diseases (BLB and Bacterial leaf streak) are described in Table 6.2.

Table 6.2 Influence of weather on disease outbreak

S. No.	Disease	Weather Conditions
1.	Blast	Cloudy skies, frequent rain and drizzles for 6-7 days favour development and severity. Blast spores are present throughout the year in the tropics. Damage to dryland rice is more severe than wetland. The disease progresses in low (22-23 °C) night temperatures more than 95% RH, 10 hours of leaf wetness period for 3-4 days.
2.	Sheath Blight	Occurs both in tropics and temperate regions. High temperature (28-32 °C) and humidity (90%) for 7-8 days increase the severity of sheath blight. Disease incidence increases as the plants grow older. Also frequent rains favour disease development. More common during rainy season than in dry season in tropics.
3.	Brown Leaf Spot	Common in poorly drained soils. The disease spreads by wind blown spores. Delayed onset of SW monsoon, RH more than 90% for 7-8 days favours the spread and development.

Table 6.2 contd...

S. No.	Disease	Weather Conditions
4.	Bacterial Leaf Blight(BLB)	High (25-30 ^{o}C) temperature, humidity during crop growth for 7-8 days increase the incidence. This disease is reported to have reduced Asia's annual rice production by as much as 60%. This is known as monsoon disease and the incidence and severity is influenced by rainfall and number of rainy days and heavy winds.
5.	Bacterial Leaf Streak	High (30-34 °C) temperature and high (85-90%) humidity for 8 - 9 days favour disease development. The bacteria are spread by wind and rain. The spread occurs through rain and irrigation water.

Symptoms of Damage of Major Rice Diseases

- **Blast (*Magnaporthe grisea*) (*Anamorph pyricularia oryzae*)**

  Small greyish dots on leaves enlarge into spindle shaped spots with brown margins. When numerous spots occur on leaves it results in the death and drying up of the plant. The node blast symptoms appear as black patches on infected nodes and infected node die. Panicle or neck blast results in improper grain filling and also chaffey ear heads similar to the damage of white ears by stem borer (Fig.6.7).

- **Sheath Blight (*Rhizoctonia solani*)**

  The disease generally appears at maximum tillering stage first as irregularly elongated lesions on leaf sheath. Later, lesions become bleached with an irregular purple brown border. When several lesions are continuous it results in blighting of sheaths (Fig.6.8).

- **Brown Leaf Spot (*Helminthosporium Oryzae*)**

  Spots appear on leaves as circular to oval reddish brown lesions surrounded by gold halo. On glumes the disease appears as black or brown spots (Fig.6.9).

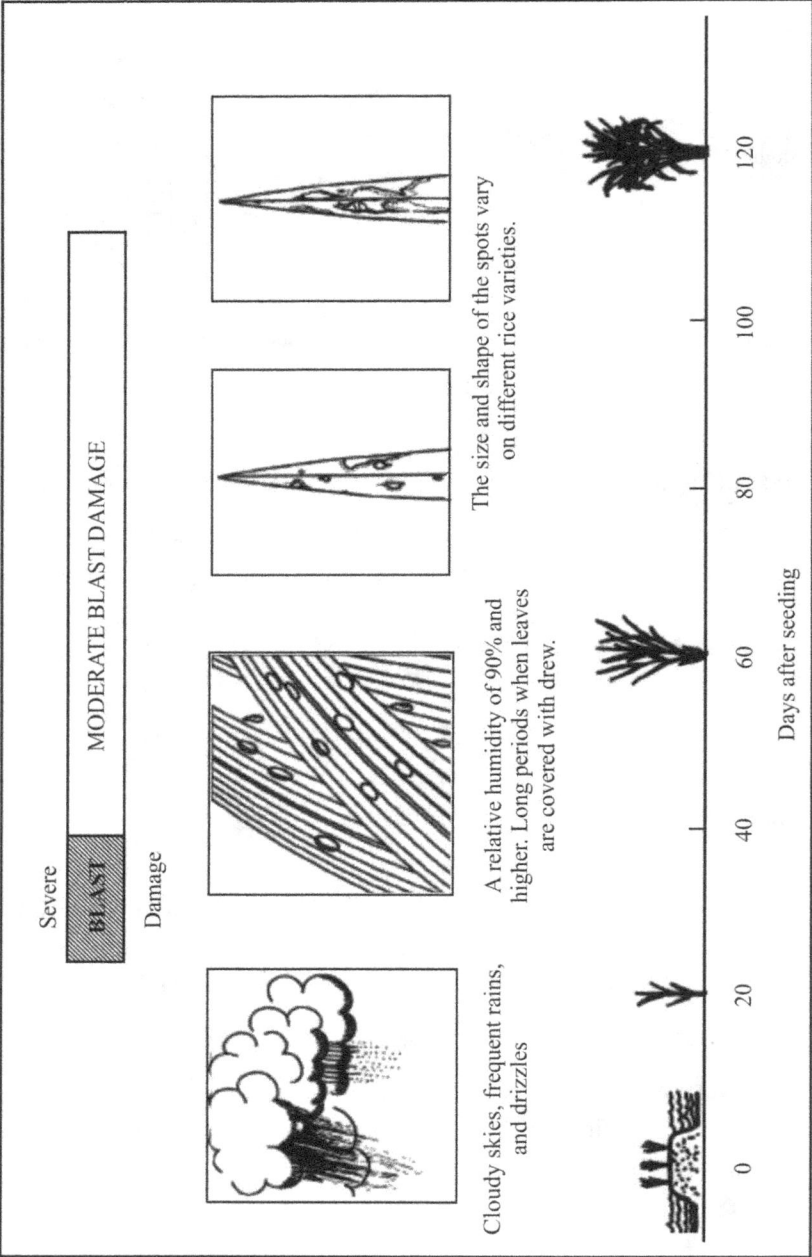

Fig. 6.7 Rice blast

- **Bacterial Leaf Blight (BLB)** (*Xanthomonas campestrics Pvoryzae*)

  Leaf margins become yellow. Drying and death of leaf occurs due to blighting of the leaf (Fig.6.10).

- **Bacterial leaf streak** (*Xanthomonas campestris* **of sp.translucens**)

  The disease is observed late in the season on leaves as narrow streaks which turn olive brown and results in premature drying (Fig.6.11).

Integrated rice disease management

- Cultivate tolerant varieties against key diseases
- Use disease free seed
- Apply need based chemicals to control all diseases
- Monitor the field continuously to maintain field sanitation
- In endemic areas of blast disease, adopt seed dressing with pyroquilon or triclyclazoic
- To reduce or control "Sheath blight" reduce or delay top dressing of nitrogenous fertilizers and apply in 2-3 split doses
- To control all diseases more so "Brown spot" apply mancozeb during early morning (avoid spray during flowering)
- To control BLB skip top dressing of nitrogenous fertilizer
- Use only the recommended dose of fungicides
- Adopt clean cultivation
- Monitor initial incidence of diseases through survey and surveillance program.

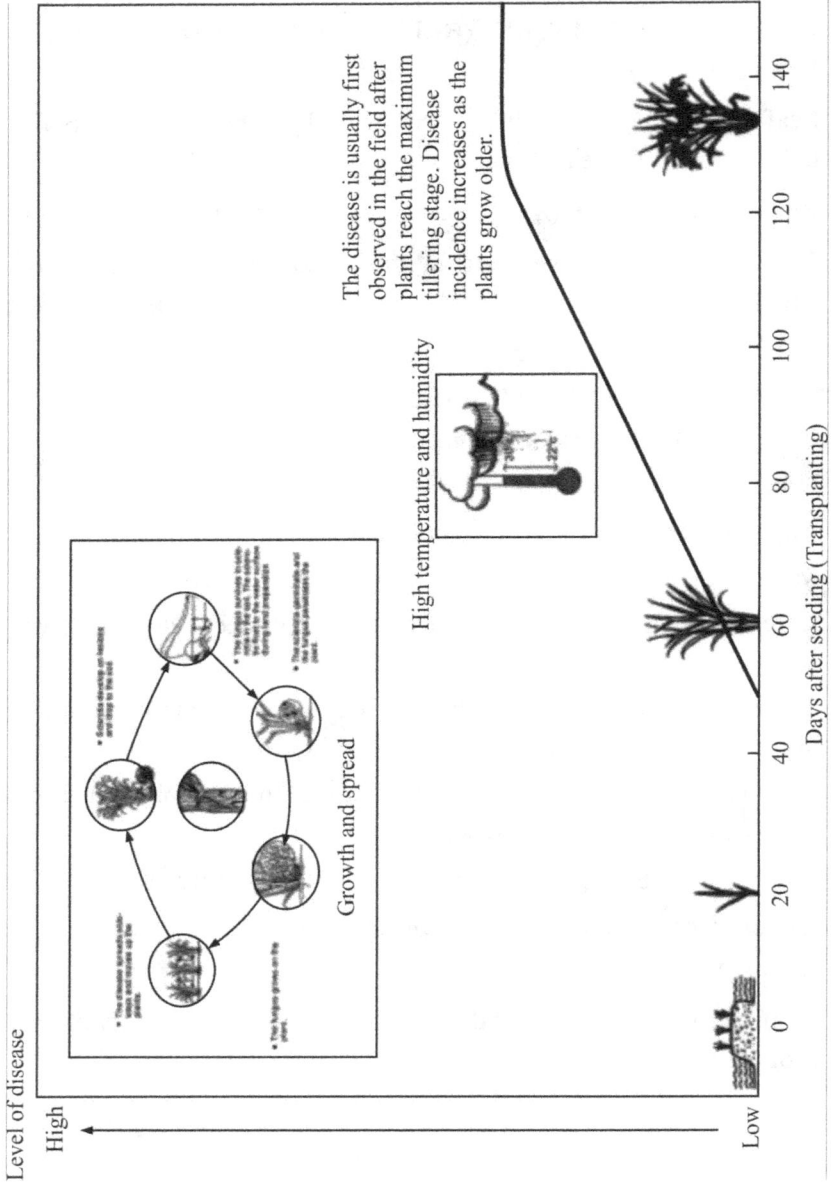

The disease is usually first observed in the field after plants reach the maximum tillering stage. Disease incidence increases as the plants grow older.

High temperature and humidity

Growth and spread

Level of disease

High

Low

Days after seeding (Transplanting)

Fig. 6.8 Rice sheath blight

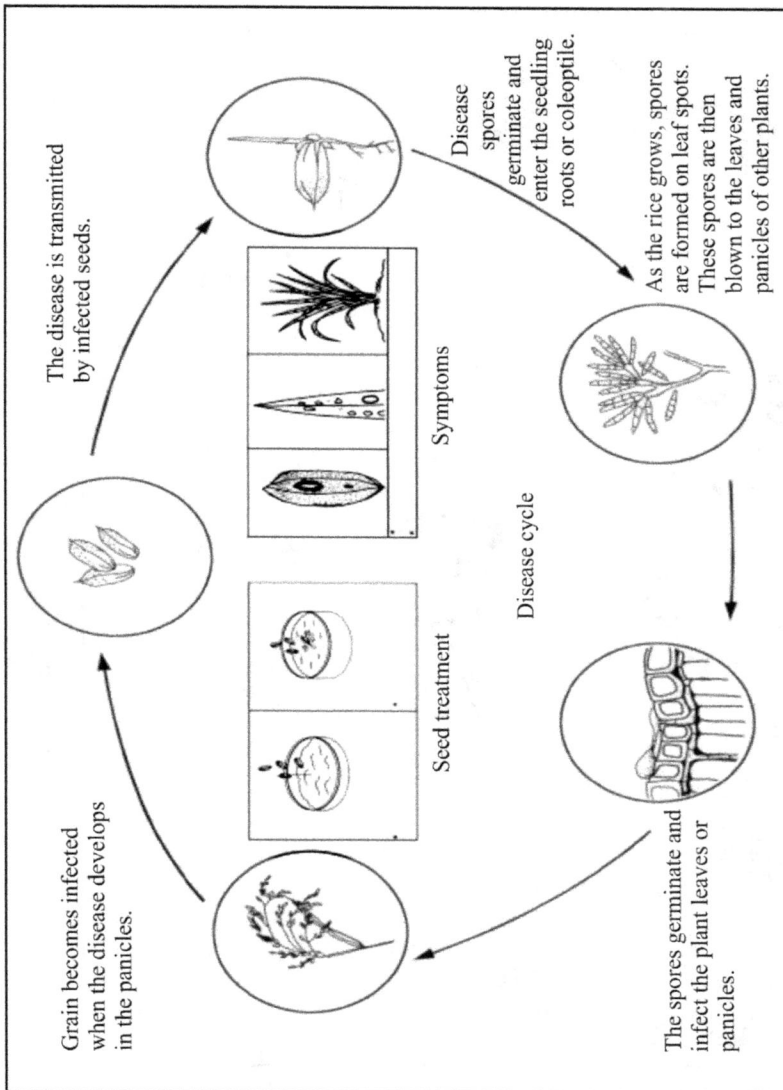

Fig. 6.9 Rice brown leaf spot

The disease is transmitted by infected seeds.

Disease spores germinate and enter the seedling roots or coleoptile.

As the rice grows, spores are formed on leaf spots. These spores are then blown to the leaves and panicles of other plants.

The spores germinate and infect the plant leaves or panicles.

Grain becomes infected when the disease develops in the panicles.

Symptoms

Disease cycle

Seed treatment

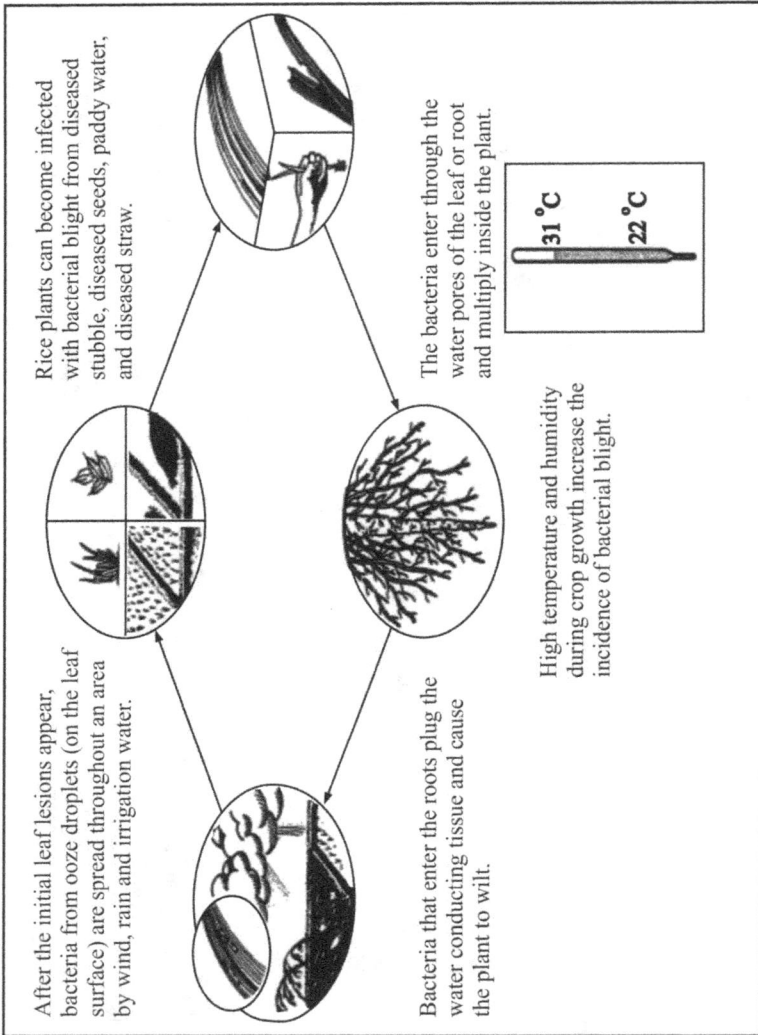

Rice plants can become infected with bacterial blight from diseased stubble, diseased seeds, paddy water, and diseased straw.

The bacteria enter through the water pores of the leaf or root and multiply inside the plant.

After the initial leaf lesions appear, bacteria from ooze droplets (on the leaf surface) are spread throughout an area by wind, rain and irrigation water.

Bacteria that enter the roots plug the water conducting tissue and cause the plant to wilt.

High temperature and humidity during crop growth increase the incidence of bacterial blight.

31 °C

22 °C

Fig. 6.10 Bacterial leaf blight

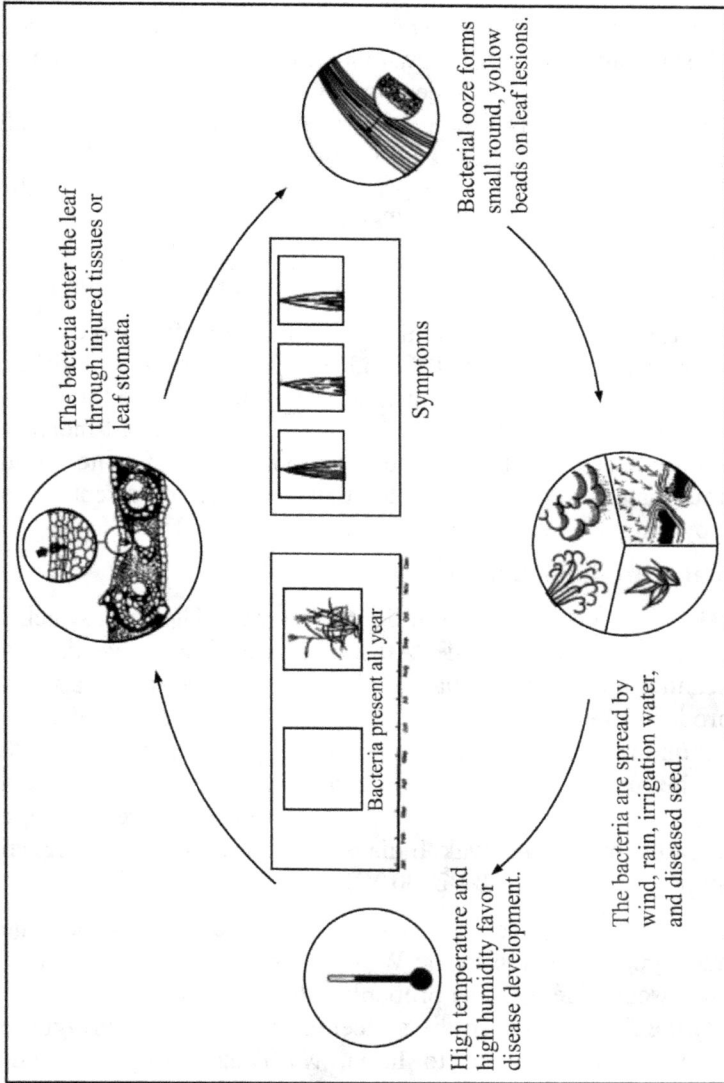

Fig. 6.11 Rice bacterial leaf streak

6.2.2   Groundnut (*Arachis hypogaea L.*)

6.2.2.1 *Preamble*

The groundnut or peanut (*Arachis hypogaea* L.) is one of the important grain legume crops predominantly grown in the tropical and the semi arid tropical countries as a principal source of edible oil and vegetable protein. Groundnuts are cultivated between 40 °N to 40 °S of the equator although yields vary enormously from about 3000 kg pods ha^{-1} in the USA to 1500 kg ha^{-1} in South America and Europe and <800 kg ha^{-1} in the developing world of Asia and Africa. Temperature extremes, especially hot air and hot soil temperatures, together with water deficits are the major environmental factors influencing the productivity of groundnuts.

Groundnut crops grown in the semi arid tropics are often exposed to soil and air temperatures as hot as 40^0C and $>35^0$C, respectively. Further, with the present trend of global warming, temperatures are likely to become hotter, perhaps by as much as 3^0 to 4^0C. This predicted increase in air temperature would further reduce the productivity of major food crops. For groundnut, it has been predicted that increases in current mean temperatures by 2^0 to 3^0C would reduce the seed yield by 23-36%.

Global Area and Production Scenario

The cultivated groundnut originated in South America. The term Arachis is derived from the Greek word "Arachos", meaning a weed, and hypogaea, meaning underground chamber i.e., in botanical terms, a weed with fruits produced below the soil surface. The earliest archaeological records of groundnuts in cultivation are from Peru, dated 3750-3900 years before the present (YBP). Groundnuts were widely dispersed through South and Central America by the time europeans reached the continent, probably by the Arawak Indians and there is archaeological evidence from Mexico, dated 1300-2200 YBP.

After European contact, groundnuts were dispersed world wide. The Peruvian runner type was taken to the Western Pacific, China, Southeast Asia and Madagascar. The Spanish probably introduced the virginia type to Mexico, *via* the Philippines, in the sixteenth century. The Portuguese then took it to Africa and later to India, *via* Brazil. Virginia types apparently reached the Southeast USA with the trade. Substantial secondary diversity in Africa and Asia, the types they found and their locations supported these various conjectures regarding dispersal.

The world groundnut (in shell) harvested area was 23.5 million ha with a total production of 30 million MT. The world's average productivity was about 1274 kg ha^{-1}. It is cultivated in close to 90 countries and requires tropical, subtropical, or warm temperate climates for optimum production. Groundnut is therefore an oilseed crop on a global scale. It grows best in regions, which lie between 40^0N and 40^0S . The most productive soils are light, friable and well-drained sandy loams (for ease of harvesting), but, the crop will also grow in heavier soils. Groundnuts are predominantly grown in the developing countries of Asia and Africa. About 90% of the total world production comes from this region and about 60% of production comes from the semi arid tropics. Roughly two thirds of this is used for oil, making it the second most important source of vegetable oil after soybean (*Glycine max* (L.) Merr.).

Uses

- Groundnuts are an important food item in Semi Arid Tropics (SAT) and are a subsistence food crop throughout the tropics
- The crop is mainly grown for the kernels, and the edible oil and meal derived from them and the vegetative residue (haulms)
- Groundnut kernels typically contain 47-53% oil and 25-36% protein
- Groundnuts are used in various forms in countries, which include groundnut oil, roasted and salted groundnut, boiled or raw groundnut or as paste popularly known as groundnut (or peanut) butter
- The tender leaves are used in certain parts of West Africa as the vegetable in soups
- Groundnut oil is the most important product of the crop, which is used for domestic and industrial purposes
- Groundnut oil is the cheapest and most extensively used vegetable oil in India
- It is used mainly for cooking, for margarine and vegetable ghee, salads, for deep-frying, for shortening in pastries and bread, for pharmaceutical and cosmetic products, as a lubricant and emulsion for insecticides and as a fuel for diesel engines
- The press cake containing 40-50% protein is used mainly as a high protein livestock feed and as a fertilizer
- The dry pericarp of the mature pods (known as shells or husks) are used for fuel, as a soil conditioner, filler in fertilizers and feeds, or are processed as substitute for cork or hardboard or composting with the aid of lignin decomposing bacteria
- The foliage of the crop also serves as silage and forage.

6. 2.2.2 *Effect of weather elements on Groundnut*

Effect of Solar Radiation

- Groundnut responds to full light intensities. Clear, cloudless days are advantageous for maximum photosynthesis and high yields can be obtained on such days
- Low light intensity prior to onset of flowering slows down vegetative growth. During rapid growth low light intensity increases height and length of stems, but, decreases leaf weight and flowering
- Low light intensities suppress development and growth of reproductive branches and consequently decrease the total flower production
- In early flowering stage causes abortion of flowers and at pegging stage significantly reduced peg number and pod weight are recorded at low light intensities
- Low light intensity during pod filling and maturity stages slightly decreases number and weight of mature pods but significantly increase % of shriveled kernels
- Shading before onset of flowering slowed vegetative growth. Shading in early flowering stage appeared to affect distribution of pods around mainstem. Shading after peak flowering interrupted in filling resulting in a significant reduction in percentage of extra large kernels
- Solar radiation detoxify groundnut oil contaminated with aflatoxin
- Highly significant negative linear relationship exists between night temperature and radiation use efficiency
- The radiation use efficiency is negatively associated with canopy extinction coefficient
- Solarisation during hot summer will provide a sufficient level of suppression on root-knot nematodes.

Effect of Temperature

- Temperature plays an important role in all aspects of plant growth and development. It is a major environmental factor that determines the rate of groundnut crop development. Heat stress and water stress are the major environmental factors limiting pod yields in groundnut. Groundnuts grown in the SAT are often exposed short episodes of temperatures or continuous periods of air and soil temperature >35 °C. Groundnuts are susceptible to both hot air and hot soil temperatures because of aerial flowering and subterranean fruiting habit of the crop

- The optimum temperature for the germination and seedling development is close to 30 °C

- Temperature significantly influences the vegetative growth of the groundnuts. The optimum temperature range for leaf and stem growth is 28° to 30 °C

- Under field conditions in Zimbabwe groundnut crop growth rate, leaf area and total dry matter production were maximal at a site with mean daily maximum and minimum temperature of 30° and 17 °C respectively

- The rate of crop photosynthesis was remarkably conservative over a range of mean air temperature from 19° to 30 °C and groundnuts attain their maximum leaf apparent photosynthesis at 30 °C and this was reduced by 25% at 40 °C

- Research has also shown that groundnuts accumulate less total biomass whenever the night temperatures are consistently at or below 16 °C

- Flowering is more sensitive to heat stress than the other phases of reproductive development. Studies in controlled environments have shown that heat stress and water deficits at flowering result in the largest reduction in pod yields in both Spanish and Virginia types, suggesting that this stage of development is more sensitive to hot temperature than the other post flowering stages of reproductive development

- Experimental results showed that three day temperatures (30°, 32° and 35 °C) with a constant night temperature of 22 °C it was found that there were no significant effects of temperature on time to first flowering or on the total number of flowers produced

- It was found that there were significant increases in the rate of flower production following the episode of heat stress and that the cumulative number of flowers produced at hot temperature (>35 °C) were greater than those produced at the optimum temperature (28 °C)

- The optimum day/night temperature for number of developing pegs ranges from 25°/25°C to 32°/23°C

- The onset of pegging and the development of pegs are especially sensitive to hot temperature. The optimum air temperature for pod initiation and development ranges between $24°$ to 28 °C

- It was reported that the optimum soil temperature range for podding in groundnut was from $31°$ to 33 °C and that soil temperatures above that range significantly reduced the number of pegs forming the pods

- The optimum temperature for pod growth is about 23 °C

- It was suggested that the longer the duration of flowering which culminates with the start of seed filling, the greater is the probability that this stage will coincide with stress temperatures in fluctuating field environments

- The optimum soil temperature for pod yields was in the range of 30 to 33 °C

- Soil and therefore root temperatures in tropical and subtropical regions are often in the range of $35°$ to $40°C$ and are detrimental to nodule formation and nitrogen fixation. At these hot temperatures, the groundnut root biomass is reduced and the roots are thin, unbranched and with very few root hairs and so produce fewer nodules

- It was reported that temperatures of 30 °C and 35 °C significantly reduced the nitrogenase activity of groundnut root nodules as compared to those at a temperature of 25 °C

- The sensitivity of groundnuts to heat stress could be correlated with the sensitivity of the nitrogen fixing ability and growth of strains of *Bradyrhizobium* to hot soil temperature. However, there has been very little work done to identify groundnut rhizobia able to effectively fix nitrogen at temperature extremes and their suitability under field conditions.

Effect of Wind

- High wind speeds (78 km per hour) results in drying up of top layer of the soil and do not support germination. It is advisable to go for deeper sowing at early stages in areas where high winds and low relative humidity persist

- A frost free gentle wind and moderate relative humidity during growing period is essential.

Effect of Rainfall

- This is the most important weather parameter which not only influences vegetative but also reproductive stages

- Erratic behavior of rainfall i.e., low amount that too highly variable coupled with unfavorable conditions (light soils with low water holding capacity) are the main causes of low yields

- The ideal pattern of rainfall distribution for groundnut is pre sowing 80-120 millimeters, at sowing 100-120 millimetres, flowering to peg penetration 200 millimeters and pod development and pod maturation 200 millimeters

- A temporary stress of 15-20 days after the stand establishment is better to check the excessive vegetative growth and to promote flowering

- Prolonged stress reduces the uptake of NPK, delays flowering, reduction in number of flowers, flowering period, flower opening and pollen viability.

6.2.2.3 *Influence of Weather on occurrence of Groundnut Pests and Diseases*

S. No.	Name of the pest/disease	Favourable weather
1	Leaf eating caterpillar, leaf webber, root grub, jassids, thrips	Frequent rain and dry spell, high rain, high RH, cloudiness, low temperature
2.	Tikka leaf spot	High RH, low temperature, cloudiness

6.2.3 Sugarcane (*Saccharum officinarum (L.,)*)

6.2.3.1 *Preamble*

- Sugarcane *Saccharum officinarum* (L.,) is an old energy source for human beings and more recently, a replacement of fossil fuel for motor vehicles

- It was first grown in South-East Asia and Western India. From 327 B.C. It had been an important crop in the Indian Sub-continent

- It was introduced to Egypt (647 A.D.) and in spain (655 AD) and also other countries and its cultivation is extended to nearly all tropical and sub tropical regions

- At present sugarcane growing countries of the world lay between the latitude 36.7° North and 31.0° South of the equator extending from tropical to sub tropical zones.

Area and Distribution

- At present sugarcane occupies an area of 20.42 m. ha with a total production of 1333 million metric tonnes

- Sugarcane area and productivity differ widely from county to country

- Brazil has the highest area (5.343 m. ha) while Australia has the highest productivity (85.1t/ha).

Uses

- Sugarcane is a renewable, natural agricultural resource, because it provides sugar besides bio fuel, fibre, fertilizer and byproducts/co-products with ecological sustainability

- Sugarcane juice is used for making white sugar, brown sugar (khandsari) Jaggery (Gur) and ethanol

- The main byproducts of sugar industry are biogas and molasses

- Molases is the chief byproduct and the main raw material for alcohol based industries.

- Excess bagasse is being used as raw material in the paper industry

- Besides, co-generation of power using bagasse as fuel is considered feasible in most sugar mills.

6.2.3.2   *Effect of Weather Elements on Sugarcane*

- Tropical and semi arid crop and short day plant

- Where cold winds do not blow and high humid prevail, then the crop yields high

- Root temperature preferred is < 27 °C

- Optimum air temperature is 30-32 °C coupled with uninterrupted sunshine increases sucrose content

- Diurnal range of temperature must be optimum i.e., day temperature must be as optimum as possible and night temperature must be relatively less. Example: 32 °C day and 22 °C night temperature is most optimum

- The winter temperature shall not be less than 12 °C (diurnal i.e., either day or night).

Management Technique

- If temperatures are low i.e., around 12 °C, then irrigate the crop during evening and next day morning

- Apply nitrogenous fertilizer
- In India it is grown from 8° to nearly 33 °N and mostly from sea level to 1500 m MSL
- Ideal temperatures for growth and development

Carbon assimilation 30 °C; Tillering 33.3 – 34.4 °C; Root growth 36 °C (soil); Shoot growth 36 °C (soil); Sugar synthesis 30 °C; Sugar transport 30-35 °C

Yield of Sugarcane–Weather

- Mean annual temperature is positively correlated with cane yield
- Standard Deviation (SD) of annual temperature is negatively correlated with cane yield.

Management Techniques

- Planting season shall coinside with relatively warmer period
- Harvesting season shall co-inside with cool and dry period with less humidity and lower minimum temperature (early morning/ before sunrise)
- Drought is the single most important factor limiting sugarcane productivity. If it occurs at critical stages, yield reduction will be high. Therefore, ensure irrigation at all critical stages.

Thermotherapy

- It is an excellent method to eradicate or reduce the primary inoculums present in seed cane
- Moist Hot Air Therapy (MHAT) effectively eliminates grassey shoot disease, ratoon stunting, red rot, smut and wilt when treated at 54 °C for 2.5 hours at relative humidity of 99%
- Hot Water treatment (HWT) can successfully check redrot, smut, ratoon stunting, grassey shoot diseases and mossaic.

6.3   Farmers

6.3.1   Roving Seminars on Weather Climate and Farmers – Special Reference to Rice Crop

6.3.1.1 *Preamble*

- Weather and climate are some of the biggest risk factors impacting on farming performance and management. Extreme weather and climate events such as severe droughts, floods and temperature shocks often strongly impede sustainable farming development particularly in the tropics and sub tropics. Factors such as climate variability and change contribute to the vulnerability of individual farmers and rural communities. This also particularly impacts on regional and world food security. Recent weather and climate research efforts have demonstrated the importance of targeted forecasting and scenario analyses in increasing overall preparedness of farmers leading to substantial overall outcomes. Farmers and farming communities throughout the world have, in most instances, survived and developed by mastering the ability to adapt to widely varying weather and climatic conditions. However, the dramatic growth in human population is imposing enormous pressure on existing farm production systems. In addition, farmers are expected to manage the more insidious effects of long term climate change that is occurring at a rapid pace. Farmers have been struggling to maintain their income by continuously trying to increase yields in their production system. Such increased productivity is associated with increased economic and environmental risk as the farming systems become more vulnerable to climate variability and climate change. These existing pressures will demand the development and implementation of appropriate methods to address issues of vulnerability to weather and climate. These will be needed to assist the farmers to further develop their adaptive capacity with improved planning and better management decisions. More targeted weather and climate information can increase preparedness and lead to better economic, social and environmental outcomes for farmers. However, weather and climate forecasting is just one of many risk management tools that play an important role in farming decision making. More effective approaches to delivery of climate and weather information to farmer may need the incorporation of a more participatory, cross disciplinary approach that brings together research and development institutions, relevant disciplines, and farmers as equal partners to reap the benefits from weather and climate knowledge. Given the current concerns with climate change and its impacts on crop productivity, especially in the developing countries of the semi arid regions, there is an urgent need to sensitize the farmers about the projected climate change in their regions and the different adaptation strategies that can be considered to cope with the projected change. Examples of more general decisions that can be aided by targeted weather and climate information include strategic and tactical crop management options, agricultural commodity marketing and policy decisions about future land use for agriculture

- It is with this background that the World Meteorological Organization (WMO) is promoting the organization of a series of one day roving seminars on weather climate and farmers in different regions of the world to sensitize farmers about the weather and climate information and its applications in the operational farm management

- Also, these seminars will increase the interaction between the local farming communities and the local staff of National Meteorological and Hydrological services (NMHSs)

- The feedback is crucial for NMHSs in providing better services to agricultural community.

Objectives

The objectives of these seminars are

- To make farmers become more self reliant in dealing with weather and climate issues as the same affect agricultural production on their farms

- To secure farmer self reliance, through helping them better informed about effective weather/climate risk management by sustainable use of natural resources for agricultural production.

Organizing the Seminars

- It will be useful to collect all available information on socio economic status in general and on agriculture in particular

- Gather information on the major crops/cropping systems in these districts

- Collect the long term daily climatic data for the districts from DHM i.e., rainfall, maximum and minimum temperatures

- Analyze the climatic data for any long term trends

- Prepare fact sheets on the climate of the district and how the crops/cropping systems are influenced by the climate

- Prepare information charts for use in the roving seminars

- Develop the agenda for the seminars and train the trainers

- Organize the seminars

- The one day seminars bring together farmers of 2-3 villages to a centralized location. Criteria followed in selection of villages:

- Representative of as much possible of geographical area as possible with regard to weather and crops
- Gender equality and age
- Large, medium and small farmers in equal proportion.

6.3.1.2 *Partners and Participation by Agencies*

Roving seminars shall be organized in full co-operation with NMHSs, local agricultural extension services with the active involvement of the agricultural research personnel from a regional agricultural research station or agricultural university in a region.

6.3.1.3 *Budget for the Seminars*

The average cost of organizing the roving seminars depends on location. The amount covers the cost of hiring an appropriate location and its preparation, production of suitable training material in local language, local organizational costs including transport, tea, lunch for all participants and travel and honorarium for the resource personnel, per-deim, food and traveling allowance to farmers has to be provided. Enough money shall be earmarked for providing information on "Weather – climate – Agriculture" in the form of pamphelets, small booklets and study material.

6.3.1.4 *Organizing the Seminars on Rice*

Number of seminars organized

To achieve the above objectives 25 roving seminars were organized in India (Andhra Pradesh) Srilanka, Bangladesh and Nepal, during 2007-2013.

Farmer's attendance

The number of farmers attended each seminar ranged from 100 to 125. The women farmers were given lecture modules separately (suitable to their farm activities involving weather). Different groups of farmers participated.

Technical aspects

Lectures

Enough care and caution was taken to make lectures more interactive and promote good dialogue with farmers. Selection of villages was completed 10 days before each seminar. General climate and weather of the region and village and its influence on crops was documented in advance in the lectures.

These lectures in local language (Telugu/Simhalese/Bangladesh/ Nepalese) by the Director/Co-director/Senior Resource persons, etc., selected for seminars focused on the following aspects;

- Introduction to weather and climate
- Introduction of terminologies used in weather/climate forecasting
- Use of short term weather forecasting in agricultural operations
- Introduction to cloud and weather map
- Introduction to seasonal climate patterns
- Climate risk in production and drought alerts in different crops
- Introduction to better risk management
- Introduction to measurement of weather elements
- Planning cropping strategy and water requirements
- Video/slide show on weather/climate disasters and their management
- Display of wall posters and laminated diagrams on weather and climate.

Murthy's Daily Weather and Agriculture (MDWA)

During each seminar the farmers were shown/given the daily weather data for the last 30 days. This data was collected from the daily newspapers available in the villages where the seminar was organised after pasting same on a white sheet in front of them a day before the event. After showing this huge and valuable information on weather that is available in their own village then the farmers responded with unparalleled enthusiasm to do the same on their own for their own farm and also community benefits. Some farmers agreed to copy/write the weather information available daily on Television and Radio and transmit exchange the same with other farmers. This operational agro-meteorological tool "DVV (Dinasari Vatavaranam- Vyasayam)" involves no money because the newspapers are bought by villagers/farmers for learning and enlightening themselves on several issues. Also, in India Srilanka, Bangladesh and Nepal newspapers are very inexpensive and Television and Radio are available in all villages. Based on the trends observed (analysis of weather data) the management options and guidance is made available to the farmers within the hand outs as also the "Vyavasaya Panchangams (Agricultural Dairy)" and the book written in local language by Dr. Murthy distributed during these seminars. This concept was explained in brief in local language to all the farmers.

6.3.1.5 *Material for Banners and Pamphlets*

General Quotations for making Banners

- "Use the power of Weather – Harvest higher yields without any investments"
- "Weather is a non-monetary input in agriculture"
- "Adopt weather based agriculture – Reap tonnes of quality grains by sowing few kilograms of seed"
- "Weather has life like you, me, plants and soil"
- "Rain water soaked soil gives support for germination of innumerable types of seeds"
- "Timely rain – Weather based agriculture – Optimal grain yields"
- "Weather wise – Otherwise unwise"
- "Good rain after anthesis and pollination gives abundant yields of crops"
- "Rainwater shall be retained and soaked only in the field (do not allow run off)"
- "Summer ploughing – Best for rainfed agriculture"
- "Summer pulse crops – Farmer's money earners"
- "Ploughing field before rains – Helps controlling soil erosion"
- "Use chemicals where essential – Protect environment"
- "Plough, such that top soil is not allowed to go out of the field - Harvest good crop"
- "Use fertilizers when essential – Protect soil health and environment"
- "Adopt contour cultivation – Save soil from erosion and increase soil water retention capacity"
- "If seed is good – crop is very good"
- "Treat the seed – Improve the yield"
- "Premonsoon ploughings kill soil born disease organisms, grubs and eggs of pests"
- "Spray the chemicals against pest and diseases in the direction of winds"
- "Use protection cloths while spraying chemicals – Keep your health sound"

- "Use green manures - Improve your soil health and get better quality crop yields"
- "Use of more than recommended doses of chemicals and fertilizers results in unhealthy soil and low quality grains"
- "Every hard working farmer in the field is a special agricultural scientist".

Specific Quotations for making Banners on Rice

- Formation of alley ways in paddy fields control gall midge
- Reduction in average air temperature (4-5 days) followed by rain causes attack by stem borer
- Dry and hot winds enhance grasshoppers that cause extensive damage
- Hot and moist winds favour damage by hispa and sucking pests
- Increase in atmospheric humidity coupled with cloudy weather and less sunny days (5-6 days) enhance damage by gall midge
- Increase in atmospheric humidity (3-4 days) and no air movement increases damage by brown plant hopper
- When South-West monsoon is delayed select medium (130-135 days) or short duration (110-120 days) varieties
- Use aged seedlings (45-50 days) and plant closely with 2-3 seedlings per hill.
- Use $2/3^{rd}$ of Nitrogenous fertilizer as basal dose (before transplanting) and remaining $1/3^{rd}$ at flowering as top dressing.

Some Golden Tips

- During South West monsoon period plough the soil after a minimum of 75mm rain is received
- During South West monsoon top dress the fertilizer only when 20 mm rain is received
- When drought occurs follow inter cultivation to conserve soil moisture in crop field
- Do not spray chemical or apply fertilizers when rain is forecast in 12 hours
- Unscientific and over use of chemicals and fertilizers results in soil, air and water pollution, in addition, causes unwanted and dangerous residues in crops

- Repeat chemical spray if it rains within 3 hours
- Uncontrolled use of fertilizers results not only in elimination of friendly insects but also increases outbreak of pests and diseases
- Use castor as trap crop in cotton and groundnut and control flying insects
- Use chrysanthemum as trap crop in cotton and Bengal gram crops.
- Use Bhendi as trap crop in cotton
- Use chrysanthemum as trap crop in redgram.

Floods – Crop Protection

Short Term Measures

- Select crop varieties which are
  - Spreading type
  - More leaves
  - Withstand flooding water
  - Produce abundant strong roots
  - Flower after floods (2-3 times i.e. flushing varieties)
- Avoid poultry
- Avoid dairying
- Avoid pisciculture.

Medium and Long term measures

- Establish social forests
- Form strong field bunds with small patches of less strong (to allow strong runoff) if necessary
- Irrigation canal and tank bunds must be strengthened periodically
- Don't allow inhabitation houses for humans and animal sheds in the down stream of rivers
- Follow the advises of elders, officers of agriculture, revenue and engineering departments
- Construct proper drainage system in the villages and also for all crop fields
- Grow vetiver hedges in all the fields at proper contours
- Insure crops
- Stake the harvested crop produce in sheds / houses with enough strength to avoid floods impact.

Seed Treatment (to Control Environmental Pollution)

- Take an earthen pot of medium size. Put the seed in it. Add seed treatment material (@ 3-4 grams of captan, thiram per kilogram of seed). Cover and tie the mouth of earthen pot with a cloth. Shake the earthen pot till the seed treatment material is mixed and coated over the seed. Use the treated seed for sowing after 24 hours.

Advantages

- Controls seed born diseases
- Controls soil born diseases
- Increase in germination percentage of seeds
- Optimum initial growth of seedings
- Enough protection from sucking pests and diseases
- Reduce in environmental pollution
- Reduce cost of cultivation of crops.

Precautions for Chemical Spraying

- Carefully read the instructions on the bottle or pocket of the chemical
- Open the seal carefully and mix chemical and water as per recommended dosage
- Don't mix the chemical with hand
- Don't spray any chemical before and immediately after harvest
- Don't use sprayers which are leaking
- Spray chemical in the direction of wind
- Protect all parts of body like mouth and eyes with appropriate cover cloth and glovers
- Don't allow children and animals where the chemicals are sprayed
- Don't drink water or eat food while spraying the chemicals
- Wash hands and take head bath after chemical sprays.

6.3.1.6   *Influence of Weather on occurrence of Pests and Diseases*

S. No.	Pest/Disease	Favorable Weather
1	Stem borer	If day temperatures during October are high and evening relative humidity is also high
2	Termites	If air temperatures are high and soil temperature is less
3	Root grub	If April temperatures during February increase gradually and in May the morning relative humidity is high
4	Top shoot borer	If air temperatures during February are hot during day and relative humidity is also high
5	Red mites	If summer rains occur, then during non- rainy days these will occur
6	Yellow mites	If relative humidity is 65-75% and air temperature range is 26-29%
7	White fly	Cloudy skies, temperatures are 29-34 degrees centigrade and relative humidity is 80-90%
8	Whip smut	If summer monsoon onsets early i.e., middle of May- June and also if July rains are heavy

6.3.1.7 *Farmer – Scientist Interaction*

Several questions and problems that farmers asked answered during farmer- scientist interactions. Of them, the occurrence, damage and control of pests and diseases, agro-meteorological, agronomic issues were discussed and solutions were given. Also, the Table 6.3 contain information on specific recommendations for certain field level problems, that were asked by the farmers.

Table 6.3 Weather related problems and recommendations

S. No.	Problem	Weather factor involved	Recommendation
1.	Weak and long seedlings; elongation of leaf blade and sheath; Low	Cloudiness / shade for over 5-4 days	Grow nurseries away from trees or tall structures. Apply 1 kg extra nitrogenous fertilizer for nursery reduce seed rate

Table 6.3 contd…

S. No.	Problem	Weather factor involved	Recommendation
	dry matter weak roots		
2.	Non–uniform grain yields	Rainy days and low temperatures delay ripening. In contrast warm and sunny days shorten ripening. Therefore variation in ripening results non-uniform grain yields	Grow photo insensitive and fertilizer responsive varieties; Right varietal selection is important in such situations.
3.	Sprouting of seeds on panicles	When it rains just before harvest non dormant seeds germinate. The problem is due to selection of non-dormant variation.	Follow weather forecast at all critical stages of crop more so at harvest. Select seeds for nursery from the harvest of dry and sunny season crop
4.	Less tillers during dry season and less dry matter	Low temperatures and less net radiation for 15-20 during vegetative stage.	Follow closer spacing; Apply 50% more fertilizer than that recommended. More fertilizer increases tillers, leaf area, rate of dry matter production and more leaves and solar radiation.
5.	Low dry matter in rainy season	Plants are tall, leafy, shade each other. Cloudy weather results in taller plants. Low light above and inside the crop during rainy season results in weak plants.	Apply low amount of fertilizer. Apply "N" in 2-3 split doses
6.	Scorching of leaves after fertilizer application resulting leaf burn	Fertilizer stick on leaves, when wet due to rain or dew fall.	Do not top dress fertilizer while leaves are wet due to rain or dew
7.	No response to top dressed	Heavy rain washes away fertilizer from the field	Do not top dress when rain is impending in 12 hours

Table 6.3 contd...

S. No.	Problem	Weather factor involved	Recommendation
	fertilizer; Pale yellow leaves		
8.	Lodging of rainy season crop	Winds and rain due to cyclones, flatten the ripened crop Closer spacing results in taller plants and weaker stems, the problem is due to growing tall variety in rainy season.	Do not grow tall varieties during rainy season because taller the plant greater the tendency to lodge.

6.3.1.8 *Feed Back and Evaluation*

An evaluation study was conducted by Dr. Murthy, the author of the book to understand the outcome of the seminars on adoption of weather and climate information. A Total of 2250 farmers, over 100 experts/resource personnel who participated in the seminars (professors, scientists, extension specialists etc.) and 450 students of agricultural sciences and polytechnics were selected on random basis as respondents. An open ended interview schedule was used for collecting data, analyzed and the results are as follows:

- Weather/ climate play a vital role in all farm operations

- "Climate change" is being observed and there is an urgent need to address the impacts of climate change on agriculture

- Weather/ Climate is a "NON-MONETARY' farm input. Since, the cost of all farm inputs are increasing at an alarming rate, the use of information on weather, as non-monetary input is the need of the hour for better quality and quantitative yields of crops and animals

- 25.2 per cent of farmers, agricultural scientists and polytechnic students followed "Murthy's Daily Weather and Agriculture (MDWA)" and used "Weather" as non–monetary input in their daily agricultural operations. They were able to reduce the cost of cultivation of crops by 10%. Also, 54.1% of the farmers and students believed strongly that they were not only able to substantially reduce the cost of cultivation but also obtained 2-3% improved quality produce by following MDWA and knowledge gained in the roving seminars, but, were unable to quantify the benefit in monetary terms

- Some farmers and polytechnic students enlightend their collegues and trained them on how to use, weather as a non- monetary input in all agricultural operations. They also reported that if weather based farming is done, the cost of cultivation of crops and animals can be reduced at least by 10% and quality of the agricultural produce be improved by 2-3%

- Dr. Murthy, the author of the book, in his works on roving seminars proved that past 10 days weather is as important as 10 day forecast of weather, in developing weather based technologies and making farm management decisions

- Also, it was proved by the author that Growing Degree Days are very much useful in predicting pests and Helio Thermal Units are highly useful in predicting diseases on crops.

References

This book is meant to those students who are pursuing advanced courses in agricultural meteorology. Therefore, to avoid confusion of repeated references in text, explanation etc., all the important works utilized by the author are mentioned in this section of references. The author wishes to inform that it is impossible to be original in an attempt of this nature and he endeavored to acknowledge all sources of information and expressions of the original authors, publications, publishers, printers etc., who own the copy right. Still, if any inadvertent omissions are found the author acknowledges the same, with utmost reverence and respect to all publishers, authors etc. The author expresses his thanks to all the other authors and publishers whose books/works he frequently consulted and referred to in this work. In addition to so many lecture notes, catalogues, cyclostyled papers, IMD/WMO manuals, monographs, instruction bulletins, ICAR/ USDA publications, research notes etc., the following works have been consulted in preparing this work.

Also references may be made to: WMO *Technical Notes* 11; 21; 26; 55; 56; 83; 86; 97; 125; 126; 133; 161; 168; 179; 192; 315.

1. Agrometeorological Services in India. A Publication of Agricultural Meteorological Division, IMD, GOI, MOES, Pune.

2. Association of Agrometeorologists 2001. Abstracts and Souvenir of national seminar on agrometeorological research for sustainable agricultural production. 27-28 September, 2001, GAU, Anand, India.

3. Biggelaar, C.D., Lal-R., Wiebe, K., Hari Eshwaran. D., Breneman, V and Reich, P. 2004. Global impact of soil erosion on productivity : II. Effects on crop yield and production over time Advances in Agronomy. Volume 81., Academic press, pp. 49-89.

4. Boggess, W.G and C.B. Amerling. 1983. A bioeconomic simulation analysis of irrigation environments. *S.J. Agric. Econ.* 15:85-91.

5. Boote, K.J., Jones, J.W., Hoogenboom, G., Wilkerson, G.G and Jagtap, S.S. 1989. PNUTGRO VI.0, Peanut crop growth simulation model, user's guide. Florida Agricultural Experiment Station, Journal No. 8420. University of Florida, Gainesville, Florida, USA.

6. Central Soil and Water Conservation Research and Training Institute (CSWCRTI) Dehradun N. India (2009) : Salient Research achievements. http://www. CSWCRTI.org.

7. Chakravarthy N.V.K and Gautam R.D. 2002, "Forewarning mustard aphid" NATP mission mode project on "Development of weather based forewarning systems for crop pests and diseases" (F.No.27 (27) / 99 NATP / MM – 111 – 17) program. Dept. of Agril. Physics, IARI, New Delhi.

8. Crawford, G.W and G.-A. Lee. 2003. Agricultural Origins in the Korean Peninsula. Antiquity. 77(295): 87-95.

9. Cross, H.Z. and Zuber, M.S. 1972. Prediction of flowering dates in maize based on different methods of estimating thermal units. *Agronomy Journal*. 64 : 351-355.

10. Das, H.P. 1999. Management and mitigation of adverse effects of drought phenomenon. In : Sinha DK, Rahim MB (eds) Natural disasters – some issues and concerns. Natural Disasters Management Cell, Visva Bharati, Shantiniketan, Calcutta, India. pp 87-103.

11. Das, H.P. 2003. Incidence, prediction, monitoring and mitigation measures of tropical cyclones and storm surges. In agrometeorology related to Extreme Events. WMO No.943, World Meteorological Organisation, Geneva, pp.

12. Das, H.P. 2004. Extreme events with particular reference to flood and heavy rainfall and their impact on agriculture. *Intromet*-2004. pp 233-234.

13. De Datta, S.K. 1981. Principles and practices of rice production, John wiley de sons, Inc Newyork. SBI91-R5D38 633-18 80-28 941

14. De Wit, C.T and Goudriaan, J. 1978. Simulation of assimilation, respiration and transpiration of crops. Simulation monograp. PUDOC, Wageningen, The Netherlands.

15. Directorate of Rice Research (DRR) 1994-2005. Progress Report of All India Coordinated Rice improvement programme, Kharif 1993-2004. Directorate of Rice Research, Hyderabad, Andhra Pradesh, India.

16. DRR (Directorate of Rice Research). 2004. Annual Report for 2003-04, p., Hyderabad.

17. DRR (Directorate of Rice Research). 2005. Annual Report for 2004-05, p. Directorate of Rice Research, Hyderabad

18. FAO 1999. New concepts and approaches to land management in the tropics with emphasis on steeplands, FAO Soils Bulletin No.75, Food and Agriculture Organisation, Rome.

19. Gilmore, E.C. Jr. and Rogers, J.S. 1958. Heat Units as a methods of measuring maturity in corn., Agronomy Journal. 50 : 611-636.

20. Gommes, R and Petrassi, F. 1996. Rainfall variability and drought in Sub-Saharan Africa since 1960. FAO Agrometeorology Series Working Paper No.9, Food and Agriculture Organisation, Rome.

21. Goudriaan, J. 1977. Crop micrometeorology: a simulation study. Simulation monograph. PUDOC. Wageningen, The Netherlands.

22. Hanyana, S. 2001. Soil erosion threatens farm lands of Saharan Africa. The earth times. http : // forests. org, networked by Ecological internet, inc., info @ ecologicalinternet.org

23. Hess, U. 2007. Weather index insurance for coping with risus in Agricultural production, Chapter 22, Same as 377-396.

24. Hoogenboon, G. 2000. Contribution of agrometerology to the simulation of crop production and its applications. Agril. and for Meteorol. 103 : 137-157.

25. Hundal S.S. 1999. Practical manual on introductory agricultural meteorology, Exercise No.12. Forecasting crop stages using GDD, Department of Agricultural Meteorology, PAU, Ludhiana, India, 36-37 pp.

26. Indian council of Agricultural Research 2008. Hand book of Agriculture Chapter 9 Land utilization pages. 204-253.

27. Jahn, G.C, Khiev, B., Pol, C., Chhorn, N., Pheng,, S and Preap, V. 2001. Developing sustainable pest management for rice in Cambodia. Pp.243-258, In S.Suthipradit, C. Kuntha, S.Lorlowhakarn, and J. Rakngan (eds.) "Sustainable Agriculture: Possibility and Direction" Proceedings of the 2nd Asia-Pacific Conference on Sustainable Agriculture 18-20 October 1999, Phitsanulok, Thailand. Bangkok (Thailand): National Science and Technology Development Agency. p. 386.

28. Jeyaseelan, A.T. and Chandrasekar, K. 2002. Satellite based identification for updation of Drought prone area in India. ISPRS-TC-VII. International Symposium on Resource and Environmental Monitoring, Hyderabad.

29. John Wiley and Sons Ltd 2009. Earth surface processes and land reforms volume 34 issue 7, pages 969-980, published on line 13 March, 2009.

30. Leung, L.K.P., Peter, G., Cox, Gary, C., Jahn and Robert Nugent. 2002. Evaluating rodent management with Cambodian rice farmers. Cambodian journal of Agriculture, Vol.5, pp21-26.

31. Maracchi, G.V., Perarnand and Kleschenko, A.D. 2000. Applications of geographical information systems and remote sensing in agrometeorology. Agricultural For...Meteorology 103:119-136.

32. McMaster, G.S. 1995. Another wheat (Triticum spp.) model ? Progress and applications in crop modeling, Rivista di Agronomia 27, 264-272.

33. McMaster, G.S., Simkar, D.E., 1988. Estimation and evaluation of winter wheat phenology in the central great plains Agirl. For Meterol. 43, 1-8.

34. McMaster, G.S and Wilhelm, W.W. 1997. Growing degree days one equation, two interpretations. Agricultural and Forest Meteorology. 87 : 291-300.

35. Monteith, J.L. and Unsworth, M.H. 2007. (3rd Ed.). Principles of environmental physics. Edward Arnold, London, in print.

36. Monteith, J.L., 2000. Agricultural meteorology: Evolution and application. Agri. For. Meteorol. 103(2000) : 5-9.

37. Motha R.P and Murthy, V.R.K., 2007. "Agrometeorological services to cope with risks and uncertainties" Chapter 25. WMO-IMD International workshop on "agro-meteorological risk management" New Delhi, 23 –27, October, 2006. ISBN 978-3-540-72744-6 Springer. Managing weather and climate risks in Agriculture. Edited Sivakumar MVK and Mothar R.P, ISBN 978-3-540-727446- Springer, Berlin, Heidelberg, New York, pp 377-404.

38. Motha R.P., Sivakumar, M.V.K and Bernardi, M. 2006. Strengthening operational agrometeorlogical services at national level. Proceedings of the Inter-regional workshop, March 22-26, 2004, Manila Philippines.

39. Motha, R. and Baier, W. 2005. 'Impacts of present and future climate variability on agriculture and forestry in the temperate regions : North America', *Clim. Change* **70 :** 137-164.

40. Murthy, V.R.K. 2010. Roving Seminars on weather climate and farmers in India and Srilanka. Paper and poster presentation in commission for Agricultural meteorology of WMO, at Belo Horizonte, Brazil in 14-16, July 2010.

41. Murthy V.R.K. 2003. The role of crop growth models in Agricultural production. Training workshop on Satellite Remote Sensing and GIS applications in Agricultural Meteorology. Dehradun. July 7-11, 2003.

42. Murthy V.R.K. and Stigter, C.J. 2003. Stigter's diagnostic conceptual framework for generation and transfer of agricultural meteorological services and information for end users. Paper in : Agrometeorology in the new millennium – perspectives and challenges. Proceedings of the Second National Seminar of the Association of Agrometeorologists in India, Ludhiana, October 26-28, 2003.

43. Murthy, V.R.K and Reddy, T.Y. 2009. The weather and climate (The sound and power) as non-monetary inputs in agriculture for under developed and developing countries. Paper and poster presented in world climate conference-3, Genera, Switzerland, August 27-Sept 2, 2009.

44. Murthy, V.R.K. 1996. Terminology on agricultural meteorology. Srivenkateswara Publishers, Ashok Nagar, Hyderabad, A.P., India, pp 125.

45. Murthy, V.R.K. 2002. Basic principles of agricultural meteorology, BS publications, 4-4-309, Giriraj lane, Sultan Bazar, Hyderabad, 435-462.

46. Murthy V.R.K. 2007. A Report on "Weather Climate – Farmers", WMO, Geneva, Switzerland.

47. Murthy, V.R.K. 2007. B "Weather based fore warning of rice (*Oriza Sativa* L.) major pests and disease under System of Rice Intensification (SRI) project presented to Department of Science and Technology, Government of India, Technology Bhavan, New Delhi – 110 016.

48. Murthy, V.R.K., Mohammed, S.K., Prasad, P.V.V and Satyanarayana, V. 2002. Resource capture mechanisms – an aid to promote nursery growth in paddy for higher yields in winter. Symposium of Association of Agrometeorologists, Anand. October 26-28, 2002.

49. Murthy, V.R.K. 1995. Practical manual on agricultural meteorology, Kalyani Publishers 1/1, Rajendernagar, Ludhiana, 86 pp.

50. Murthy, V.R.K. 1999. Studies on the influence of macro and micro meteorological factors on growth and yield of soybean. Unpublished Ph.D thesis submitted to ANGRAU, Hyderabad, India.

51. Murthy, V.R.K. 2002. Basic Principles of Agricultural Meteorology, B.S. Publications, 4-4-309, Giriraj Lane, Sultan Bazar, Hyderabad-95, Andhra Pradesh, India,   260 pp.

52. Nene, Y.L. (Ed).2005. Agricultural Heritage of Asia: proceedings of the International Conference, 6-8 December 2004, Asian Agri-History Foundation, Secunderabad-500009, Andhra Pradesh, India.

53. Nuttonson, M.Y. 1955. An Instt., Crop Ecol, Washington D.C, 150.

54. Penning de Vries, F.W.T. 1977. Evaluation of simulation models in agriculture and biology: conclusions of a workshop.   Agricultural Systems, 2:99-105.

55. Pruess, K.P. 1983, Day – degree methods for pest management, Environ. Ent. 12, 613-619.

56. Reddy, J.S. 2002. Dryland agriculture: An agroclimatological and agrometeorological perspective. B.S. Publications 4-4-309, Giriraj Lane, Sultan Bazar, Hyderabad- 500 095, ISBN 81-7800-022-09.

57. Riksen. M., Brouwer. F and Graaf. J.D. 2003. Soil conservation policy measures to control wind erosion in northwestern Europe. Wind erosion in Europe. 52: 3-4, pp. 309-326.

58. Salinger, J., Sivakumar, M.V.K. and.Motha, R.P 2005. Increasing climate variability and change, reducing the vulnerability of agriculture and forestry. Springer, P.O.Box 322, 3300 AH Dordrecht, The Netherlands ISBN 1-4020-3354-0.

59. Sankara Reddy, G.H and Yellamanda Reddy, T. 1995. Efficient use of irrigation water. Kalyani publishers, Ludhiana, India.

60. Sinclair, T.R. 1986. Water and nitrogen limitations in soybean grain production. I. Model development. Field Crops Res. 15 : 125-141.

61. Singh, B.G and Aruna Kumari, 2005. Terminology on plant physiology, BS publication, 4-4-309, Girirajlane, Sultan Bazar, Hyderabad. ISBN 81-7800-089-X.

62. Sivakumar, M.V.K and Glinni, A.F. 2000. Applications of crop growth models in the semi-arid regions. WMO, bis Avenue de la Paix, 1211 Geneva, 2 Switzerland.

63. Sivakumar, M.V.K, Gommes, R and Baier, W. 2000. Agrometerology and sustainable agriculture, Agril. and for meteorology. 103(2000) : 11-26.

64. Sivakumar, M.V.K., Zöbisch, M.A., Koala, S. and Maukonen, T. 1998. Wind erosion in Africa and West Asia: problems and control strategies. ICARDA, Aleppo, 198 pp.

65. Sivakumar, M.V.K., Roy, P.S., Harmsen, K and Saha, S.K. 2004. Satellite Remote Sensing and GIS Applications in Agricultural Meteorology. Proceedings of the Training Workshop in Dehradun, India. AGM-8, WMO/TD-No. 1182, WMO, Geneva.

66. Sivakumar, M.V.K., Motha, R.P. and Das, H.P. 2000. Natural disasters and extreme events in agriculture. Impacts and Mitigation. Salinger Verlag Berlin Heidelberg 2005, ISBN-10 3-540- 90-4.

67. Stigter, C.J., Zheng Dawei., Xurong, M. and Onyewoto, L.O.Z. 2005. 'Using traditional methods and indigenous technologies for coping with climate variability'. *Clim. Change*, **70** : 255-271.

68. Thapliyal, P.K., Pal, P.K. and Gupta, A. 2004. Use of passive microwave radiometer data for monitoring drought conditions. Intromet-2004. pp 47-49.

69. Van Dam, C. 1999. La Tenencia de la Tierra en America Latina. El Estado del Arte de la discussion en la region inicitiva global tierra. Territorios y Derechos de Acceso Santiago, IUCN Regional office South America.

70. Van Keulen, H. and Wolf, J (ed). 1986. Modelling of agricultural production: weather, soilis and crops. Simulation Monographs, PUDOC, Wageningen, The Netherlands.

71. Wang Chunlin, Liu Jinluan, Zhou Guoyi, 2003. Research on real-time cold-disaster watching and prediction in Guangdong Province based on GIS technology. Quarterly Journal of Applied Meteorology, Vol. 14(4). Pp 487-495.

72. Wieringa, J. and E. Rudel, 2002. Station exposure metadata needed for judging and improving quality of observations of wind, temperature and other parameters. Paper 2.2 in WMO Technical Conference on Meteorological and Environmental Instruments and Methods of Observation (TECO-2002). [Also available on CD-ROM].

73. Wiley and Sons Ltd. 2009. Earth and science processes published online 2009, volume 34, issue: pages 969-980.

74. WMO 1997. Extreme agrometeorological events. CAgM Report No.73, TD NO.836, World Meteorological Organisation, Geneva.

75. WMO, 1980. Compendium of Lecture notes for training Class IV agrometeorological personnel (by A.V. Todorov). WMO-No. 593.

76. WMO, 1981. Meteorological aspects of the utilization of solar radiation as an energy source. WMO No. 557, 298 pp.

77. WMO, 1984 (2nd Ed.). Compendium of lecture notes for training Class IV Meteorological Personnel (by B.J. Retallack), Vol.II-Meteorology, WMO No.266, Geneva.

78. WMO, 1994. Guide to hydrological practices: data acquisition and processing, analysis, forecasting and other applications. WMO No. 168, Geneva.

79. WMO, 2001. Lecture notes for training agricultural meteorological personnel (by J. Wieringa and J. Lomas), WMO No.551, Geneva.

80. WMO, 2003. Guidelines on climate metadata and homogenization (by E. Aguilar, I. Auer, M. Brunet, T.C. Peterson and J. Wieringa). WMO-TD No. 1186 (WCDMP-No. 53).

81. WMO, 2006 (7th Ed.). Guide to meteorological instruments and observing practices, WMO-No.8, in print. Available at the WMO/CIMO website, Geneva.

82. WMO, 2007 (3rd Ed.). Guide to climatological practices, WMO-No.100, Geneva, in preparation.

83. World Meteorological Organisation at a glance. A WMO Publication; Geneva, Switzerland.

84. World Disasters Report, 2003. International Federation of Red Cross and Red Crescent Societies, Geneva.

85. Yellamanda Reddy, T and Sankara Reddy, G.H. 1992. Principles of Agronomy. Kalyani publishers, Ludhiana, India.

86. Zobeck, T.M., Sterk, G., Funk, R., Rajot, J.L., Stout, J.E and Van Pelt, R.S. 2003. Measurement and data analysis methods for field-scale wind erosion studies and model validation. Earth Surf. Process. Landfs. 28: 1163-1188.

Index

www.ingramcontent.com/pod-product-compliance
Lightning Source LLC
Chambersburg PA
CBHW050656190326
41458CB00008B/2590